Mainstreaming Biodiversity into Renewable Power Infrastructure

OECD

BETTER POLICIES FOR BETTER LIVES

Please cite this publication as:
OECD (2024), *Mainstreaming Biodiversity into Renewable Power Infrastructure*, OECD Publishing, Paris, https://doi.org/10.1787/357ac474-en.

ISBN 978-92-64-96579-9 (print)
ISBN 978-92-64-91256-4 (pdf)
ISBN 978-92-64-75329-7 (HTML)
ISBN 978-92-64-67580-3 (epub)

Photo credits: Cover © foxbat/Shutterstock.com

Corrigenda to OECD publications may be found on line at: www.oecd.org/about/publishing/corrigenda.htm.

Foreword

Countries are grappling with two interconnected challenges: climate change and biodiversity loss. Both have severe implications for human health, societal well-being and the economy, necessitating urgent and ambitious action. While efforts to tackle biodiversity loss and climate change are often mutually reinforcing, conflicts and trade-offs may arise. This is evident in the expansion of renewable power.

To achieve the temperature goals of the Paris Agreement, the world must electrify energy end-uses and rapidly scale up low-emissions sources of electricity. In IEA's Net Zero Emissions by 2050 scenario consistent with the 1.5 degrees Celsius target, renewable power capacity increases three-fold by 2030 and more than eight-fold by 2050. While reducing climate-related pressures on biodiversity, renewable power expansion could exacerbate other pressures. Some renewable power projects adversely affect biodiversity, for example, through habitat loss, direct species mortality and disturbance. Such impacts can accumulate across projects and over time with potentially significant consequences.

The transition to low-emissions power systems must align not only with the Paris Agreement but also the new Kunming-Montreal Global Biodiversity Framework. This requires an integrated, strategic approach to sector and project planning that systematically considers biodiversity from the outset. Failure to account for biodiversity has not only negatively affected nature, it has resulted in renewable power projects being delayed or cancelled, thereby undermining the transition.

This report, Mainstreaming Biodiversity into Renewable Power Infrastructure, underscores the need to simultaneously increase renewable power and protect biodiversity. In addition to comprehensively reviewing the impacts of solar power, wind power and power lines on biodiversity, the report presents good practices and tools for mainstreaming biodiversity into power sector planning and policy. It also discusses the suite of policy instruments governments can use to encourage the industry to mitigate adverse biodiversity impacts and promote positive outcomes.

With most renewable power infrastructure yet to be deployed, we have a considerable opportunity to develop electricity systems that deliver better outcomes for both climate and biodiversity. I believe this report provides government planners, regulators and environmental policy makers with valuable guidance on how to protect our species and ecosystems while scaling up renewable power.

Acknowledgements

This report was developed by the OECD Environment Directorate, directed by Jo Tyndall. The report was authored by Edward Perry, Biodiversity Policy Analyst in the Climate, Biodiversity and Water Division (CBW), under the guidance of Walid Oueslati, Head of CBW Division and Katia Karousakis, Biodiversity Programme Lead, CBW.

The author is grateful for the substantive inputs provided by Coline Pouille (OECD Environment Directorate), and the comments received from the following colleagues: Geraldine Ang, Andrew Brown, Joseph Cordonnier, Jane Ellis, Katia Karousakis, Sophie Lavaud, Virginie Marchal, Hugh Miller, Mariana Mirabile, Daniel Nachtigall, Mikaela Rambali, Marcia Rocha, Klas Wetterberg (Environment Directorate), Juan Garin, Benjamin Michel (Directorate for Financial and Enterprise Affairs), Dominique Blaquier, Juan Casado Asensio (Development Co-operation Directorate), Robert Addison (Directorate for Public Governance) and Jinsun Lim (International Energy Agency). The author is also grateful for the guidance provided by external experts Leon Bennun (The Biodiversity Consultancy), Robert Adamczyk (EBRD) and Samir Whittaker (Ørsted).

The Secretariat thanks delegates from the OECD Working Party on Biodiversity, Water and Ecosystems for their comments on earlier drafts and the oversight provided by the Environment Policy Committee. The Secretariat is also grateful to the countries that completed a project questionnaire: Bulgaria, Brazil, Colombia, France, Latvia, Norway, Peru and the United States.

Table of contents

FIGURES

TABLES

BOXES

List of acronyms and abbreviations

CBA	Cost-benefit analysis
CBD	Convention on Biological Diversity
CEA	Cumulative effects assessment
CO_2	Carbon dioxide
CRM	Collision risk model
CSP	Concentrated solar power
EC	European Commission
EIA	Environmental impact assessment
EJ	Exajoules
ELIS	Environmental labelling and information schemes
EMF	Electromagnetic fields
EPR	Extended producer responsibility
EU	European Union
EUR	Euros
GIS	Geographic information systems
GW	Gigawatt
Ha	Hectare
IAS	Invasive alien species
IEA	International Energy Agency
IFC	International Finance Corporation
KBA	Key biodiversity area
KW	Kilowatt
LT-LEDS	Long-term low greenhouse gas emission development strategies
MCDA	Multi-criteria decision analysis
MNE	Multinational enterprises
MSB	Migratory soaring bird
MW (MWh)	Megawatt (megawatt hours)
NBSAP	National biodiversity strategies and action plan
NDC	Nationally determined contribution
NGO	Non-governmental organisation
ODA	Official development assistance
PA	Protected area
PMSG	Permanent magnet synchronous generator
PPA	Power purchase agreement
PV	Photovoltaics
RBC	Responsible business conduct
REZ	Renewable energy zones
RoW	Right of way
SDGs	Sustainable development goals
SEA	Strategic environmental assessment

TW (TWh)	Terawatt (terawatt hours)
UK	United Kingdom
US (USD)	United States (US dollars)
VECs	Valued environmental and social components

Executive Summary

The world faces simultaneous climate and biodiversity crises, with profound implications for human health, well-being and the economy. Biodiversity loss and climate change are closely entwined. Healthy ecosystems regulate the climate and provide services such as flood protection that support societal adaptation; their degradation worsens climate change. Climate change is the fastest growing driver of biodiversity loss.

Significant synergies exist between climate and biodiversity actions, but also potential conflicts and trade-offs. This is evident in renewable power expansion. Electrifying energy use and increasing renewable power are pivotal to limiting global average temperature increase to 1.5°C and, therefore, to addressing global biodiversity loss. However, without careful planning and management renewable power infrastructure's expanding footprint could, itself, drive declines in biodiversity. This, in turn, could undermine ecosystems' resilience to climate change, their capacity to sequester carbon and their contribution to societal adaptation.

The challenge for countries is to rapidly increase renewable power without compromising their commitments to halt and reverse biodiversity loss under the Kunming-Montreal Global Biodiversity Framework. This demands an integrated approach to harness synergies, minimise trade-offs and avoid unintended consequences. It requires governments to systematically consider and address both climate and biodiversity objectives throughout electricity planning and policy.

This report reviews the evidence for biodiversity impacts of renewable power infrastructure, focussing on solar power (photovoltaics and concentrated solar power), wind power (onshore and offshore) and power lines. It identifies good practices for mainstreaming biodiversity into power sector planning and policy. While the report underscores the need to address adverse impacts from sourcing minerals for renewable power infrastructure, it is primarily geared towards understanding and addressing impacts from constructing and operating renewable power infrastructure.

Biodiversity impacts from wind power, solar power and power lines

Renewable power infrastructure can impact biodiversity in various ways, including direct species mortality (e.g., from collision or electrocution); habitat loss and degradation; habitat fragmentation and barrier effects on species movement; habitat alteration; behavioural and physiological changes; and ecosystem services impacts. These impacts can accumulate across projects, time and political boundaries, potentially leading to declines in species' populations and ecosystem integrity. The nature and extent of impacts depend on the type of infrastructure (e.g., solar photovoltaics or concentrated solar power) and its design, where and how critical mineral inputs are mined and processed, where the infrastructure is located and how it is constructed, operated, maintained and decommissioned.

While the evidence base for renewable power impacts on biodiversity has grown substantially in recent years, it is incomplete and uneven across technologies, species, ecosystems and geography. For example, the impacts of wind on terrestrial ecosystems are better understood than those on marine ecosystems. Similarly, the risk factors and consequences for birds are more comprehensively studied than for other species groups. Most detailed data and insights come from Europe and North America; significant gaps exist in developing countries and other geographies where most new renewable energy development is projected to occur. Key knowledge gaps and uncertainties common to all renewable power include population-level effects, how species impacts can have knock-on effects on ecological communities and ecosystem services, cumulative impacts (e.g. for migratory species), and indirect impacts.

Mainstreaming biodiversity into power sector planning

Strategic planning is crucial for developing power systems that deliver better outcomes for both climate and nature. The selection of electricity generation technologies and where they are deployed have significant implications for biodiversity. By considering biodiversity early in the planning process, governments can greatly reduce the risks to biodiversity posed by renewable power projects.

Incorporating biodiversity data into energy planning models can help identify electricity capacity expansion options that balance cost, carbon emissions and biodiversity protection. The creation of renewable energy zones, designated based on biodiversity considerations, can steer renewable power projects away from areas where they pose high risk to biodiversity (e.g., key biodiversity areas; migratory routes), towards low-risk areas. When evaluating power sector policies, plans and programmes, it is imperative for governments to assess cumulative impacts on biodiversity, for example through strategic environmental assessments. Policy appraisal tools, such as multi-criteria and cost-benefit analysis, should seek to integrate biodiversity and ecosystem service values.

A key strategy for low-risk siting of renewables is to optimise the use of rooftops and other existing infrastructure for solar panels. Another strategy is to capitalise on the abundance of converted lands such as brownfields and abandoned agricultural areas. Co-locating renewable power with other economic activities can alleviate land and sea-use pressure while capitalising on potential synergies across social, economic and environmental objectives. Examples include situating solar panels amidst wind turbines or integrating power infrastructure with agriculture or aquaculture.

Effective co-ordination across ministries and levels of government is critical for addressing policy synergies and trade-offs. Opportunities exist to enhance cross-border collaboration, for example, through joint spatial planning, knowledge exchange and development co-operation to strengthen countries' capacities to mainstream biodiversity. Connecting grids across national or state borders could increase opportunities for low-cost, low-risk siting of renewables, but countries must assess and mitigate potential adverse impacts from extending transmission infrastructure.

Mainstreaming biodiversity in electricity planning not only mitigates adverse biodiversity impacts but can also provide certainty to project developers, investors and regulators. It can reduce the risk of project delays and failures, permitting time, and the project costs associated with biodiversity mitigation measures. Ultimately, biodiversity mainstreaming can contribute to a swift and sustainable transition to low-emissions power.

Policy instruments for aligning renewable power expansion with biodiversity

The right mix of policy instruments can ensure power companies mitigate adverse biodiversity impacts in accordance with the mitigation hierarchy (avoid, minimise, restore and offset), and seek positive biodiversity outcomes. An effective policy response may comprise regulatory instruments (e.g., requirements for environmental impact assessments, monitoring and data-sharing), economic instruments (e.g., grants for research and development; biodiversity offsets), and voluntary or information instruments (e.g., integrating biodiversity criteria in project tenders; ecolabelling). Scope exists to increase the use and effectiveness of these instruments to promote a transition to low-emissions power that benefits both climate and nature.

1 Recommendations and policy options

This chapter summarises the report's key recommendations and policy options for mainstreaming biodiversity into renewable power infrastructure. It intends to support government planners, regulators and environmental policy makers to plan and deliver low-emissions electricity transitions that are biodiversity-aligned.

Electrifying the economy and investing in renewable power are essential to decarbonising energy and achieving key global goals, including the temperature goals of the Paris Agreement and various Sustainable Development Goals (SDGs), such as SDG 3: Good Health and Well-being and SDG 7: Universal Access to Affordable and Clean Energy. As climate change significantly contributes to nature's decline, expanding renewable power could also contribute to the Kunming-Montreal Global Biodiversity Framework's mission to halt and reverse biodiversity loss. Exceeding a global temperature rise of 1.5 degrees Celsius above pre-industrial levels could substantially worsen biodiversity loss.

Though renewable power expansion reduces climate-related pressures on biodiversity, it brings its own risks. Renewable power projects can lead to direct species mortality (e.g., from avian or bat collision with infrastructure); habitat loss, degradation and fragmentation; changes to ecosystem services; and other impacts. These impacts can accumulate across multiple projects. Without careful management, renewable power developments could drive significant population declines for sensitive species and disrupt ecosystem function. Furthermore, by degrading intact ecosystems, poorly planned renewable projects could inadvertently lead to avoidable greenhouse gas emissions and undermine ecosystems' capacities to support societal resilience.

Various solutions exist for mitigating the adverse impacts of renewable power. Examples include careful siting (and micro-siting) of infrastructure away from ecologically sensitive areas, designing power lines to minimise electrocution risk and increasing the cut-in speed of wind turbines to reduce the risk of bat collisions. Such solutions are being tested and refined, as experience and evidence increase. Digital technologies such as machine learning and artificial intelligence (AI) are providing new opportunities for the industry to monitor and cost-effectively mitigate impacts on biodiversity (e.g., wind turbine shutdown-on-demand mechanisms based on automated monitoring of collision-sensitive species). Through strategic planning and targeted policies, governments can help scale up such solutions to ensure that renewable power expansion does not compromise biodiversity goals.

This report aims to help governments achieve their biodiversity objectives, while swiftly transitioning to low-emissions electricity systems. It explores opportunities for harnessing synergies between renewable power expansion and biodiversity objectives, managing trade-offs and avoiding unintended consequences. The report's objectives are: 1) to synthesise evidence on the biodiversity impacts from infrastructure for renewably sourced electricity generation and from power distribution and transmission infrastructure; and 2) to share insights and good practices for integrating biodiversity considerations into electricity sector planning and the policies governing the development of renewable power infrastructure and power lines. The analysis focuses on infrastructure for solar power, wind power and electricity transmission and distribution. Wind and solar power are the focus because they are expanding faster than other technologies and have greater potential to be scaled up globally than other sources of low-emissions electricity.

This chapter summarises the key recommendations and policy options of the report. It intends to guide planners, regulators and environmental policy makers as they pursue a biodiversity-aligned transition to low-emissions electricity systems. The recommendations focus on mainstreaming biodiversity into planning and policy. While mainstreaming biodiversity into renewable energy financing (e.g., through sustainable finance taxonomies, biodiversity safeguards and nature-related disclosure) is also important, it is outside the report's scope and therefore not addressed in the recommendations.

While the report is based primarily on an analysis of wind energy, solar energy and power lines, many of the recommendations hold for other renewable and non-renewable power projects. It should be stressed that regulations for fossil fuel energy projects – and enforcement of these regulations – must be at least as stringent given the significant threat fossil fuels pose to biodiversity as well as human health and well-being. The key recommendations and policy options are provided below.

- **Scale up efforts to hold global average temperature increase to 1.5°C, pursuing low-energy demand pathways that deliver benefits for climate, biodiversity and other well-being objectives.**
  - Adopt ambitious low-emissions development strategies and policies that integrate climate, biodiversity, energy and broader well-being objectives.
  - Leverage the full range of demand-side mitigation measures, including technological and social innovations, to reduce energy demand by improving energy efficiency and changing consumer behaviour.
  - Apply a systems approach to the design (re-design) of energy end-use systems (e.g., transport, food) so that they require less energy and materials.
- **Consider biodiversity impacts when selecting among power sector technologies and capacity expansion options.**
  - Integrate spatially explicit biodiversity data into power system modelling to identify capacity expansion options that are relatively low cost, low emissions and low risk for biodiversity.
  - Evaluate the cumulative biodiversity impacts of capacity expansion options through appropriate environmental assessments.
  - Integrate ecosystem service values and biodiversity-related measures into cost-benefit analysis or multi-criteria decision analysis tools used to appraise technology choices and capacity expansion options.
- **Prioritise renewable power deployment in areas of low ecological sensitivity and avoid the most ecologically sensitive areas.**
  - Develop biodiversity-explicit spatial plans for renewable power infrastructure, stipulating no go areas and areas of low ecological risk where renewable power projects should be prioritised.
  - Ensure that siting decisions account for potential cumulative impacts.
  - Accelerate solar rooftop deployment as a low-risk option e.g., by subsidising installation of solar panels on existing roofs and mandating solar panels for public buildings and new builds.
  - Promote research and development of technologies and approaches for co-locating renewable power infrastructure with other infrastructure (e.g., integrated solar-wind power facilities; solar PV in motorway sound barriers) and activities (e.g., food production; ecosystem restoration).
  - Adapt land and sea-use regulations to facilitate co-location and the siting of renewable power in areas of low ecological risk (e.g. brownfield sites; abandoned agricultural land), while managing potential indirect impacts from activities displaced by renewable energy facilities.
- **Develop policies and guidance to ensure that power generation, transmission and distribution projects effectively mitigate adverse impacts on biodiversity.**
  - Review requirements, processes and guidance for environmental impact assessment and permitting, to promote efficiency and ensure risks to biodiversity are effectively addressed.
  - Ensure renewable power companies and utilities strictly adhere to the Mitigation Hierarchy (avoid, minimise, restore onsite and where appropriate offset) to address adverse impacts.
  - Establish "no net loss" (or "net biodiversity gain") requirements for new infrastructure projects, including power sector infrastructure, with robust metrics and methods for verification.
  - Adopt standards to promote infrastructure designs and operational practices with lower-risk to biodiversity (e.g., bird-safe power-line design; minimum cut-in-speed for wind turbines).
  - Encourage or require renewable power companies and their investors to conduct due diligence in line with the OECD's Due Diligence Guidance for Responsible Business Conduct.

 - Require post-construction monitoring and reporting by projects to ensure that recommendations generated through environmental assessments and permitting requirements are respected and to inform adaptive management.
- **Encourage positive biodiversity outcomes from power generation, transmission and distribution projects.**
  - Integrate biodiversity criteria into tenders for renewable power projects to incentivise companies to go beyond regulatory requirements.
  - Establish or endorse certification schemes with science-based criteria to encourage power sector projects to seek positive biodiversity outcomes (e.g., pollinator-friendly solar).
  - Encourage power companies to adopt ambitious biodiversity targets (e.g., net positive by 2030), a plan to achieve the targets and a methodology for assessing progress. Collaborate with power companies on proactive actions to conserve and restore nature.
- **Strengthen the quality and transparency of data on biodiversity and renewable power interactions.**
  - Support development and application of environmental sensitivity mapping tools to inform project siting decisions.
  - Develop protocols and guidelines for monitoring biodiversity impacts from renewable power. Encourage coordination across projects to evaluate and address their cumulative impacts.
  - Require sharing of data from SEA, EIA, other pre-construction surveys and post-construction monitoring. Develop open access data platforms to facilitate data sharing.
  - Support targeted research to address knowledge gaps on the impacts of renewable power on biodiversity (e.g., understudied species, ecosystems, renewable power technologies and geographies – particularly developing countries) and the effectiveness of mitigation measures.
- **Promote cross-border collaboration to mitigate the adverse biodiversity impacts of transitioning to low-emissions electricity systems.**
  - Promote collective ambition to protect biodiversity in renewable power developments across sub-national and national governments to ensure species and ecosystems are protected across their entire range and lifecycle.
  - Promote cross-border spatial planning and impact assessments, share data and information on biodiversity affected by renewable power and co-ordinate policy to better understand and address cumulative impacts on transboundary ecosystems and migratory or mobile species.
  - Harness opportunities presented by cross-border electricity trade for siting renewable power infrastructure in areas of low ecological risk, while assessing and managing adverse biodiversity impacts from transmission infrastructure.
  - Leverage official development assistance to develop partner country capacity to integrate biodiversity into energy planning and policy, support biodiversity-explicit spatial planning processes, and establish monitoring and data management systems.
- **Address upstream biodiversity (and other) adverse impacts from the sourcing and processing of minerals and the manufacturing of parts for renewable power infrastructure.**
  - Prioritise mining in areas of relatively low ecological risk and avoid sites that have particularly high biodiversity values that may be compromised by mining.
  - Promote international good practice principles in mining, ensuring full application of the Mitigation Hierarchy by companies extracting or refining minerals.
  - Promote supply chain transparency and apply due diligence guidelines to promote sustainable extraction and trade of the minerals required for the low-emissions transition.

- Pursue greater resource efficiency and material circularity for renewable power infrastructure through extended producer responsibility and other policies that promote resource productivity, material recovery, sustainable materials management and the 3Rs (i.e., reduce, reuse, recycle).

2 Biodiversity, infrastructure and the low-emissions transition: Context for action

This chapter provides the context for mainstreaming biodiversity into infrastructure for renewable power generation, transmission and distribution. It shows how the electricity sector will need to grow and transform to meet energy and climate goals and discusses the implications for biodiversity. The chapter also presents the economic case for mainstreaming biodiversity into renewable power infrastructure.

Built infrastructure is ubiquitous in modern day life. It supports myriad services such as electricity generation and delivery, mobility and accessibility, telecommunications, flood defence, water provision and waste treatment. Delivery of infrastructure services is central to sustainable development: infrastructure is the focus of Sustainable Development Goal (SDG) 9 Industry, Innovation and Infrastructure and supports many other SDGs, such as SDG 6 Clean Water and Sanitation, SDG 7 Affordable and Clean Energy, SDG 11 on Sustainable Cities and Communities and SDG 13 Climate Action. Sustainable infrastructure investment has a crucial role to play in ensuring a strong, resilient, sustainable and inclusive recovery from the COVID-19 crisis, as highlighted by the 2020 OECD Ministerial Council Statement (OECD Ministerial Council, 2020[1]).

A key sustainability challenge is to ensure infrastructure investment is harmonious with nature. Biodiversity[1] and the ecosystem services it provides underpin human well-being, livelihoods and economic prosperity (Dasgupta, 2021[2]; OECD, 2021[3]). Halting and reversing biodiversity loss is a global objective reflected in SDG 14: Life Below Water and SDG 15: Life Above Land, and the Convention on Biological Diversity's (CBD) Kunming-Montreal Global Biodiversity Framework (Box 2.1). However, infrastructure investment in pursuit of other policy objectives often drives biodiversity loss. Infrastructure has contributed (directly or indirectly) to all five of the key pressures on biodiversity: land and sea-use change, over-exploitation of natural resources, climate change, pollution and the spread of invasive alien species (Balvanera et al., 2019[4]).

To mitigate infrastructure's negative impacts on biodiversity, decision makers must systematically consider and address infrastructure's dependencies and impacts on biodiversity at every stage of policy, planning, programme and project cycles. This is a process referred to as biodiversity mainstreaming in the CBD. While effective mainstreaming requires collective action from an array of actors (national and subnational governments, business, the finance sector, civil society organisations and citizens), national governments play a pivotal role in driving mainstreaming through planning, regulation, economic incentives and procurement.

Box 2.1. The Kunming-Montreal Global Biodiversity Framework

At the 15th Conference of the Parties to the Convention on Biological Diversity (CBD) in 2022, signatories to the CBD adopted the Kunming-Montreal Global Biodiversity Framework to "catalyze, enable and galvanize urgent and transformative action [...] to halt and reverse biodiversity loss".

The framework's vision is a world of living in harmony with nature where "by 2050, biodiversity is valued, conserved, restored and wisely used, maintaining ecosystem services, sustaining a healthy planet and delivering benefits essential for all people." It's mission for 2030 is "[to] take urgent action to halt and reverse biodiversity loss to put nature on a path to recovery for the benefit of people and planet by conserving and sustainably using biodiversity and by ensuring the fair and equitable sharing of benefits from the use of genetic resources, while providing the necessary means of implementation."

The Kunming-Montreal Global Biodiversity Framework has four long-term goals for 2050 related to the 2050 Vision for biodiversity and 23 action-oriented global targets for 2030. The four main goals are:

Goal A The integrity, connectivity and resilience of all ecosystems are maintained, enhanced, or restored, substantially increasing the area of natural ecosystems by 2050; Human induced extinction of known threatened species is halted, and, by 2050, the extinction rate and risk of all species are reduced tenfold, and the abundance of native wild species is increased [...].

Goal B Biodiversity is sustainably used and managed and nature's contributions to people, including ecosystem functions and services, are valued, maintained and enhanced, with those currently in decline being restored, supporting the achievement of sustainable development for the benefit of present and future generations by 2050.

Goal C The monetary and non-monetary benefits from the utilization of genetic resources and digital sequence information on genetic resources, and of traditional knowledge associated with genetic resources, as applicable, are shared fairly and equitably [...]

Goal D Adequate means of implementation, including financial resources, capacity-building, technical and scientific cooperation, and access to and transfer of technology to fully implement the Kunming-Montreal Global Biodiversity Framework are secured [...].

Source: CBD (2022[5]), Kunming-Montreal Global Biodiversity Framework, www.cbd.int/doc/decisions/cop-15/cop-15-dec-04-en.pdf.

Bolstering efforts to mainstream biodiversity into infrastructure over the coming decade will be particularly important owing to two inter-related trends in infrastructure. First, demand for infrastructure will continue to grow and, without adequate consideration of biodiversity, lead to further biodiversity loss. This demand growth is driven by macro trends such as global population growth, urbanisation and rising global gross domestic product, although the extent of demand growth will also be determined by societal and political choices. Second, the architecture and characteristics of infrastructure networks are transforming as new technologies develop and countries pursue sustainable development objectives. This transformation can bring new opportunities and challenges for biodiversity.

One sector where these trends are prominent is the electricity (power) sector. Electricity comprises an increasing share of total energy consumption. Global demand for electricity is projected to roughly double by 2050 and almost quintuple by 2100 (IPCC, 2022[6]). At the same time, the power sector is transforming. The share of renewable power (e.g., solar, wind, bioenergy) in the electricity generation mix is increasing, owing to their falling costs and climate policy (IEA, 2021[7]). Furthermore, distributed energy resources (e.g., rooftop solar) are facilitating the decentralisation of power systems (OECD, 2019[8]).

The growth and transformation of the power sector has mixed implications for biodiversity. Understanding and managing these implications will be critical to ensure that the expansion of renewable power reinforces rather than compromises efforts to halt and reverse biodiversity loss, in line with the Kunming-Montreal Global Biodiversity Framework.

2.1. Aim and scope of the report

The report aims to help governments simultaneously scale up renewable power and protect biodiversity. Its objectives are: 1) to synthesise evidence on the biodiversity impacts from infrastructure for renewably sourced electricity generation and from power lines for electricity distribution and transmission; and 2) to share insights and good practices for integrating biodiversity considerations into power sector planning and policy for renewable energy.

For an in-depth analysis of impacts and targeted policy recommendations the report focuses on solar power (photovoltaics [PV] and concentrated solar power [CSP]), wind power (onshore and offshore) and infrastructure for electricity transmission and distribution. Wind and solar power are the focus because they are expanding faster than other technologies and are set to be dominant in the global energy mix. Furthermore, the resource potential for these technologies is globally widespread. Unless otherwise specified, the terms "renewable power" and "renewable energy" are used synonymously in this report.

While the report underscores the importance of considering impacts across the full life cycle of renewable power infrastructure (e.g., mining impacts from material sourcing), it is primarily geared towards addressing the impacts arising from the construction (installation) and operation of renewable power infrastructure.

The remainder of this chapter provides an overview of the growth and transformation of the power sector, its changing spatial footprint and implications for biodiversity. It then elaborates on the economic case for mainstreaming biodiversity into renewable power infrastructure and the low-emissions transition more generally. Chapter 3 examines the evidence from biodiversity impacts resulting from solar PV, solar CSP, onshore and offshore wind and power lines. Chapter 4 explores opportunities for mainstreaming biodiversity into low-emissions pathways and power sector planning. Chapter 5 examines the policy instruments governments can use to maximise synergies and reduce trade-offs between renewable power infrastructure expansion and biodiversity.

2.2. Renewable power and the transition to low-emissions electricity systems

Global energy demand grew on average by 1.3% per year from 2010-22 (IEA, 2023[9]). A large share of this growth in demand was for electricity as countries pursued electrification. Worldwide the current share of electricity in total final energy consumption is 20% (IEA, 2023[9]). Despite increased energy supply and substantial progress in global energy access, 675 million people in developing countries remained without access to electricity in 2021 (UN, 2023[10]).

The balance of electricity generation infrastructure still favours fossil fuels but is shifting to renewables and other low-emission technologies. The global share of fossil fuels in electricity generation declined from around 67% in 2010 to 61% in 2022 (IEA, 2023[9]). Renewables are the second largest source of global electricity generation behind coal. Owing to sustained policy support and sharp cost reductions for solar PV (i.e., solar panels) and wind power, installing new renewable capacity has become the cheapest way of increasing electricity generation in most countries. As a result, growth of wind power and solar PV accounts for most of the growth in global electricity output in recent years (IEA, 2023[9]).

Mitigating climate change and ensuring universal access to affordable, reliable, sustainable and modern energy (SDG 7), will require an even greater increase in renewable power and associated transmission infrastructure. Several possible global pathways exist to achieve the goals of the Paris Agreement of "holding the increase in the global average temperature to well below 2°C above pre-industrial levels and pursuing efforts to limit the temperature increase to 1.5°C above pre-industrial levels" (IPCC, 2022[6]). These pathways vary in the use of energy resources, in energy supply and conversion technologies, in supply and end-use efficiency gains and in the extent of reliance on negative emission technologies such as bioenergy with carbon capture and storage. However, all pathways likely to limit warming to 2°C or lower include a decline in the carbon intensity of energy, owing to electrification of transportation, heating and industrial sectors and an increase in low-emissions electricity generation (IPCC, 2022[6]). Electricity supplies 48%-58% of final energy in 2050 in pathways consistent with 1.5°C with no or limited overshoot and 36%-47% of final energy in pathways that limit temperature increase to 2°C (IPCC, 2022[6]).

In the IEA's Net Zero Emissions Scenario, which is one of the pathways consistent with limiting the global temperature rise to 1.5 °C with limited overshoot, the share of electricity in final energy use jumps from 20% in 2022 to 53% in 2050 (Figure 2.1). The share of renewable energy in electricity generation globally increases from 30% in 2022 to nearly 90% in 2050 (IEA, 2023[9]). The largest growth comes from solar PV and wind (Figure 2.2). Global energy supply declines slightly by 2050 in the net zero emissions scenario due to reductions in energy intensity, but it increases in a scenario based on current and stated policies. Electricity supply increases sharply in all scenarios (Box 2.2).

Transforming the global energy system in line with the Paris Agreement goals and the SDGs requires a major reallocation of energy investments. In IPCC 1.5°C pathways, annual investment needs in low-carbon energy reach between USD 0.8 and USD 2.9 trillion globally to 2050, overtaking investments in fossil fuels by around 2025 (IPCC, 2018[11]). Most of these investments are directed at clean electricity generation, particularly solar power (USD 0.09-1 trillion/year) and wind power (USD 0.1-0.35 trillion/year). Investments for electricity transmission, distribution and storage also increase in 1.5°C pathways to support electrification of end-use sectors (IPCC, 2018[11]).

Figure 2.1. World total final energy consumption in IEA's Net Zero Emissions Scenario

Energy consumption in exajoules (EJ)

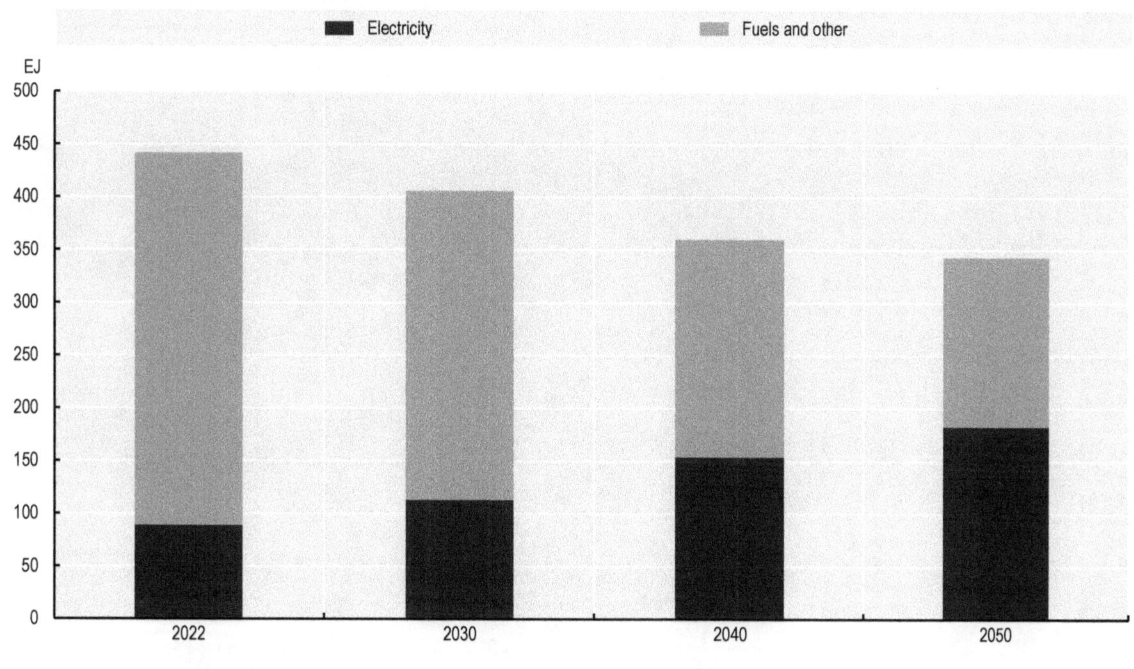

Source: IEA (2023[9]), World Energy Outlook 2023, https://iea.blob.core.windows.net/assets/42b23c45-78bc-4482-b0f9-eb826ae2da3d/WorldEnergyOutlook2023.pdf.

Figure 2.2. World electricity sector in IEA's Net Zero Emissions Scenario

Electricity generation in terawatt hours (TWh)

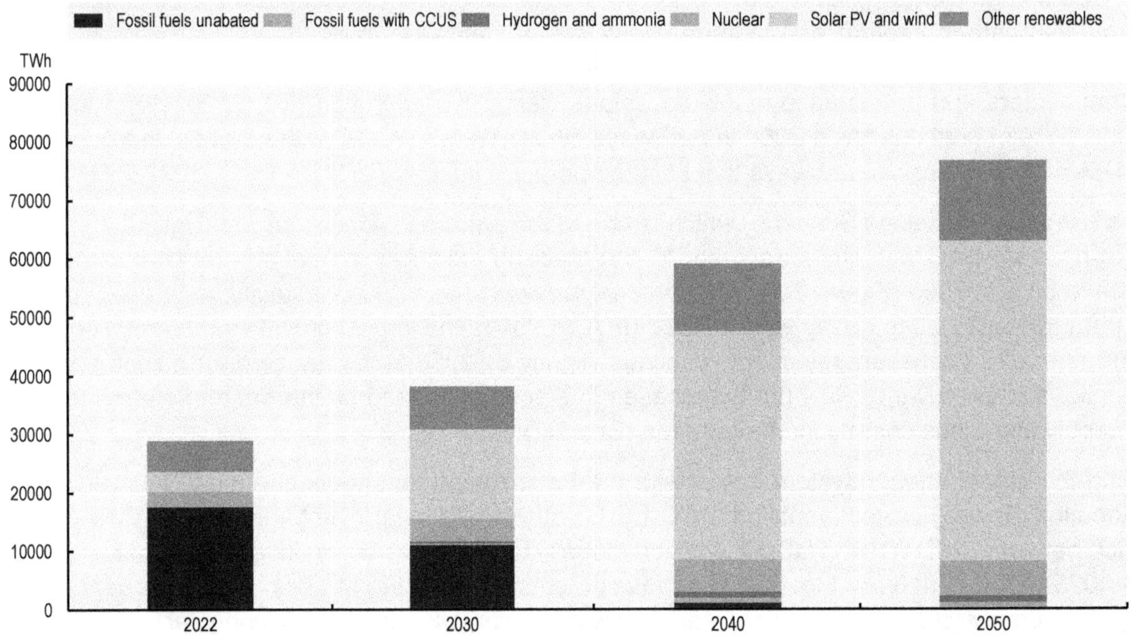

Note: Renewable energy accounts for an increasing share of electricity supply in IEA's Net Zero Emissions Scenario, reaching 89% of electricity supply in 2050. Solar and wind power expands the most, accounting for over 70% of total electricity supply by 2050.
Source: World Energy Outlook 2023, https://iea.blob.core.windows.net/assets/42b23c45-78bc-4482-b0f9-eb826ae2da3d/WorldEnergyOutlook2023.pdf.

Box 2.2. IEA scenarios for energy development

The International Energy Agency's World Energy Outlook 2023 outlines three illustrative scenarios for energy development. These scenarios and their implications for renewable energy growth are outlined below.

The IEA Stated Policies Scenario (STEPS) illustrates the consequences of existing and stated policies for the energy sector, which would result in global average surface temperature rise of 2.4°C above pre-industrial levels in 2100 (with a 50% probability). Global total final energy consumption increases 1.1% per year to 2030 and continues rising (at a slower rate) through to 2050. Global electricity demand increases by over 80% by 2050. Renewables, primarily solar PV and wind, provide 70% of global electricity generation in 2050 (up from 30% in 2022). The combined capacity of solar PV and wind triples from 2022-30 and octuples from 2022-50, reaching 16 513 GW.

The Announced Pledges Scenario (APS) assumes that all governments fully meet all the climate-related commitments they have announced, including longer term net zero emissions targets and pledges in Nationally Determined Contributions (NDCs), and commitments in related areas such as energy access. The APS is associated with a temperature rise of 1.7°C in 2100 (with a 50% probability). Global total final energy consumption increases until the mid-2020s before starting a gradual decline. Electricity generation increases 120% by 2050. Renewables, primarily solar PV and wind, provide 82% of global electricity generation in 2050. The combined capacity of solar PV and wind almost quadruples from 2022-30 and increases eleven-fold from 2022-50, reaching 21 920 GW.

The net zero emissions by 2050 (NZE) is one possible pathway for the global energy sector to achieve net-zero CO_2 emissions by 2050. It is a pathway consistent with limiting the global temperature rise to 1.5 °C above pre-industrial levels in 2100 (with at least a 50% probability) with limited overshoot, while achieving universal access to sustainable energy by 2030. Total final energy consumption declines by an annual average of 0.9% every year from now to 2050. Global electricity demand increases 150% by 2050. Renewables, primarily solar PV and wind, provide 89% of global electricity generation in 2050. The combined capacity of solar PV and wind more than quadruples from 2022-30 and increases almost thirteen-fold from 2022-50, reaching 26 369 GW.

Source: IEA (2023[9]), World Energy Outlook 2023, https://iea.blob.core.windows.net/assets/42b23c45-78bc-4482-b0f9-eb826ae2da3d/WorldEnergyOutlook2023.pdf.

2.3. Biodiversity implications of electrification and renewable power expansion

Economy-wide electrification and renewable power expansion is not only vital for achieving climate and development objectives but also biodiversity objectives. Fossil fuel combustion for energy is the primary source of anthropogenic climate change, which is the fastest growing driver of biodiversity loss. Climate change has already shifted species distribution, disrupted species interactions, and led to mismatches in the timing of migration, breeding and food supply (IPBES-IPCC, 2021[12]). Climate trends and extremes are pushing marine and terrestrial ecosystems closer to thresholds and tipping points (Harris et al., 2018[13]). Allowing global average temperature to exceed 1.5 degrees Celsius (°C) could significantly increase adverse impacts for some species and ecosystems (Smith et al., 2018[14]) (Nunez et al., 2019[15]). For example, if warming is limited to 2°C, 8% of vertebrates, 18% of insects and 16% of plants could lose at least half of their current range by 2100. However, if warming is limited to 1.5°C, this risk is halved for vertebrates and plants, and cut by two-thirds for insects (Warren et al., 2018[16]).

Furthermore, the extraction and transportation of fossil fuels, including production of charcoal in many developing countries, leads directly to land- and sea-use change (including deforestation) and pollution, and indirectly to the spread of invasive species, and illegal exploitation of natural resources (e.g. timber and wildlife) (Butt et al., 2013[17]; Harfoot et al., 2018[18]; Giam, Olden and Simberloff, 2018[19]). Coal mining, in particular, is associated with high biodiversity loss compared to other energy sources (Holland et al., 2019[20]); it can substantially alter the land surface and subsurface, resulting in a near-complete loss of terrestrial and wetland habitats for mammals and amphibians (McManamay, Vernon and Jager, 2021[21]).

While having much lower carbon footprints (UNECE, 2021[22]), renewable sources of electricity generation are not environmentally benign. They too pose a risk to biodiversity. For example, renewable power projects can result in direct mortality of flora and fauna, habitat loss and fragmentation and declines in ecosystem services (Gasparatos et al., 2017[23]; Murphy-Mariscal, Grodsky and Hernandez, 2018[24]). These and other impacts are examined in detail in Chapter 3.

Electrification and the shift towards renewables will increase the physical footprint of the power sector, owing to the greater demand for electricity and the relatively high land-use intensity (low power density) of renewable power sources. This in turn could drive land-use change (Lovering et al., 2022[25]; van Zalk and Behrens, 2018[26]). For example, a 25-80% penetration of solar in the electricity mix of the EU, India, Japan and Korea by 2050 could result in solar energy facilities alone occupying 0.5–5% of total land, contributing directly or indirectly (e.g. through displacement of agriculture and forestry) to biodiversity loss and land-based greenhouse gas emissions (van de Ven et al., 2021[27]). In the US, where energy sprawl (renewable and non-renewable) has already become the biggest driver of land-use change (Trainor, McDonald and Fargione, 2016[28]), the power sector's footprint is estimated to increase by 15 million hectares (ha) in a scenario where renewable power provides 80% of electricity generation in 2050, which is more than 50% larger than the baseline year of 2018 (van Zalk and Behrens, 2018[26]).

While the majority of electricity generation infrastructure is located on land, the spatial footprint of offshore electricity is also expanding. Under one of the European Commission's (EC) scenarios for a climate-neutral energy sector in 2050, offshore wind capacity increases to a total of 450 GW. An estimated 85% of this is from installations in the northern seas. This is the equivalent of 76 000 square kilometres (an area just under the size of Ireland) (European Commission, 2020[29]). On the one hand, harnessing more offshore energy can reduce pressure on scarce land resources; on the other hand, it may pose a risk to marine biodiversity (see Chapter 3) and conflict with other marine economic activities (e.g., fishing).

The spatial footprint of electricity distribution and transmission infrastructure will also increase (Luderer et al., 2019[30]). The growing share of electricity in final energy consumption, the increasing share of renewables in electricity supply and the need for system flexibility necessitate significant expansion of electricity grids (IEA, 2021[31]). In a scenario where temperature increase is held "well below 2°C", the annual pace of grid expansion needs to more than double in the period to 2040. Around 50% of the increase in transmission lines and 35% of the increase in distribution network lines are attributable to the increase in renewables (IEA, 2021[31]).

The growing spatial footprint of the power sector may coincide with areas of particular importance to biodiversity. One-third of areas with high potential for solar and wind energy globally, and half of areas with high potential for bioenergy, are areas with high biodiversity values (Santangeli et al., 2015[32]). Of 12 658 fully operational large scale (nominal generation capacity >10MW) onshore wind, solar PV and hydropower facilities, around 2 206 (17%) are in at least one of three important biodiversity areas: protected areas (PAs), key biodiversity areas (KBAs) and Earth's remaining wilderness areas[2] (Rehbein et al., 2020[33]). Wind power overlaps with the largest number of important conservation areas. A further 922 facilities under development are in one of these three areas; 100 of these are in strict protected areas. Combined, large-scale operational facilities and those under development (n = 3 128) overlap with 886 PAs, 749 KBAs and 40 distinct wilderness areas. Western Europe accounts for the largest number of operational sites overlapping with important biodiversity areas, while Africa and the Middle East have the largest share of their facilities that overlap (38% and 33% respectively). For the facilities under development, over half of overlapping sites are in India, Southeast Asia, South America or Africa (Rehbein et al., 2020[33]).

The spatial and ecological footprint of renewable power expansion extends beyond electricity generation, transmission and distribution infrastructure. Wind turbines, solar photovoltaics, batteries and other low-emission technologies are resource-intensive. An onshore wind plant, for example, requires nine times more mineral resources than a gas-fired power plant (IEA, 2021[31]). Low-emission technologies depend on a variety of critical minerals, such as lithium, graphite, cobalt, nickel, manganese, copper, cadmium and rare earth elements (e.g. dysprosium and neodymium used in some wind turbines; indium and tellurium used in certain photovoltaic cells) (Sovacool et al., 2020[34]; Luderer et al., 2019[30]). Demand for these minerals is expected to increase considerably, although resource efficiency may help curb demand growth for some minerals depending on rebound effects (OECD, 2019[35]; Luderer et al., 2019[30]; IEA, 2021[36]). In 2050, annual demand from energy technologies for graphite, lithium and cobalt could be nearly 500% greater than 2018 production (Hund et al., 2020[37]).

Renewable power expansion will lead to increased mining activity for critical minerals (but reduced coal mining). Current mining activity (based on 62 381 pre-operational, operational or closed sites) potentially influences 50 million km^2 of land, which is 37% of global land area excluding Antarctica. Approximately 8% of this area coincides with PAs, 7% with KBAs, and 16% with remaining wilderness areas. Most current mining areas[3] (82%) target materials required for renewable power production (Sonter et al., 2020[38]). In addition its biodiversity impacts, the rapidly growing market for critical minerals could be subject to price volatility, increasing costs due to inflationary pressures, geopolitical influence and potentially supply disruptions that could hamper the low-emissions transition.[4]

2.4. Economic case for mainstreaming biodiversity into renewable power infrastructure and low-emissions transitions

A clear economic case exists for mainstreaming biodiversity into renewable power infrastructure and the low-emissions transition. Biodiversity underpins all economic activities and human well-being. It provides critical life-supporting ecosystem services, including the provision of food and clean water, but also largely invisible services such as flood protection, carbon sequestration, nutrient cycling, water filtration and pollination. More than half of the world's gross domestic product – USD 44 trillion – is moderately or highly dependent on biodiversity (WEF, 2020[39]). Biodiversity is also fundamental to combatting climate change (Box 2.3).

Despite nature's importance, society is accumulating produced (and human) capital at the expense of natural capital (Dasgupta, 2021[2]). The population sizes of mammals, birds, fish, amphibians and reptiles have declined by an average of 69 percent since 1970, and many of the world's ecosystems are degraded (WWF, 2022[40]). Current extinction rates are tens to hundreds of times higher than the baseline rate and increasing (Diaz, S. et al., 2019[41]). The accelerating loss of biodiversity increases the risks and costs to the economy, the financial sector and society (Dasgupta, 2021[2]; OECD, 2021[3]). It undermines the provision of critical ecosystem services and nature's resilience. As ecosystems are non-linear, even small increases in pressure could lead to abrupt and irreversible ecosystem collapse, leading to economic shocks.

Box 2.3. Role of biodiversity in addressing climate change

Properly functioning ecosystems are critical for sequestering atmospheric carbon. Conserving, restoring and improving the management of forests, grasslands, wetlands and agricultural lands could deliver an estimated 23.8 gigatonnes of cumulative CO_2 emission reductions by 2030. About half of this mitigation potential represents cost-effective climate mitigation, defined as a marginal abatement cost of less than or equal to 100 USD per tonne of CO_2 by 2030. Intact ecosystems are also important for the resilience of people and infrastructure assets to landslides, flooding and other climate-related hazards; they form an effective complement or alternative to grey infrastructure. For example, mangroves in Florida are estimated to have prevented USD 1.5 billion in direct flood damages to infrastructure assets from Hurricane Irma in 2017. Maintaining or restoring the diversity of plant and animal species is fundamental for ecosystem resilience, and therefore the ability of ecosystems to continue providing mitigation and adaptation benefits in the future.

Source: Griscom et al. (2017[42]), Natural Climate Solutions, 10.1073/pnas.1710465114; OECD (2020[43]), Nature-based solutions for adapting to water-related climate risks https://dx.doi.org/10.1787/2257873d-en; Narayan, S. et al. (2019[44]), Valuing the Flood Risk Reduction Benefits of Florida's Mangroves; Loreau and De Mazancourt (2013[45]), Biodiversity and ecosystem stability a synthesis of underlying mechanisms, https://doi.org/10.1111/ele.12073; Oliver et al. (2015[46]), Declining resilience of ecosystems functions under biodiversity loss, https://doi.org/10.1038/ncomms10122.

Renewable power developments that are not strategically planned and well-managed will further pressure biodiversity and could thereby compromise efforts to achieve climate goals. Kiesecker et al. (2019[47]) estimated the required renewable energy to meet Paris Climate Agreement emission reduction targets for 109 countries[5] based on nationally determined contributions submitted by May 2016. They concluded that if CSP, solar PV and onshore wind energy facilities would be deployed where energy resources are highest, they could convert over 11 million ha of natural land (approximately the size of Bulgaria), emitting 415 million tons of carbon stored in plant biomass and soils (equivalent to about 8.6% of the combined emission reduction goals of the NDCs at the time). Avoiding land-based emissions through strategic siting could save USD 47.5-155.9 billion based on social carbon costs (Kiesecker et al., 2019[47]). A separate study estimated that direct and indirect land cover changes associated with solar energy expansion in the EU, India, Japan and Korea could cause a net carbon release of 0-50 gCO_2/kWh, depending in part on how land is managed at solar facilities (van de Ven et al., 2021[48]).

Applying biodiversity constraints to energy infrastructure siting may slightly increase system costs. For example, Wu et al. (2023[49]) estimated a 3% increase in energy system net cost in 2050 for the western United States. However, this is likely to be partially offset by the avoided costs associated with project disruptions. Biodiversity concerns have resulted in projects being delayed or cancelled owing to resistance from stakeholders (e.g. local communities and environmental NGOs) and requirements by regulators to significantly revise project plans (Dashiell, Buckley and Mulvaney, 2019[50]; Tegen et al., 2016[51]; Energy Transitions Commission, 2023[52]). For developers, this could entail unexpected costs, lost revenue and penalties if they break power purchase agreements. Facing uncertainty, investors may require a higher rate-of-return, threatening the competitiveness of renewables (Susskind et al., 2022[53]). In the US, for example, concerns over wind power siting related to wildlife, public engagement and other factors could increase costs and decrease wind capacity by 14% by 2030 and 28% by 2050 (Tegen et al., 2016[51]). Project delays can translate to higher electricity costs, affect the stability and management of electricity networks and delay the transition to a low-emissions future (Tegen et al., 2016[51]; Wind Europe, 2017[54]).

When biodiversity is insufficiently addressed in power sector planning and project development, developers may also face unexpected increases in operating and maintenance costs, for example, where turbines are required to be shutdown regularly to avoid avian or bat collision or where power lines must be repaired after wildlife collision or electrocution. Globally, wildlife-induced power shortages cost electricity suppliers an estimated USD 10 billion per year globally (Barrett, 2015[55]).

Conversely, considering biodiversity at the earliest stages of planning can facilitate a smooth and rapid deployment of renewable power while helping to minimise trade-offs with biodiversity. A study of 16 utility-scale solar projects in the US suggests that siting projects in areas of low biodiversity value can lead to a three-fold increase in permitting speed and reduce project costs associated with mitigating adverse biodiversity impacts, with overall project cost savings of 7-14% (Dashiell, Buckley and Mulvaney, 2019[50]).

The importance of ensuring renewable power and other infrastructure is sustainable – and consistent with biodiversity objectives – is reflected in several international agreements (see Table 2.1 for a summary and Annex A for further details). These agreements provide a guiding framework and impetus for countries to integrate biodiversity as they develop their domestic policy and plan the transition to low-emissions electricity systems.

Table 2.1. Summary of multilateral agreements relevant to mainstreaming biodiversity into renewable power infrastructure

Multilateral Agreement	Elements most relevant to mainstreaming biodiversity into renewable power infrastructure
Multilateral environmental agreements (MEAs)	
United Nations Framework Convention on Climate Change and the Paris Agreement	Governments committed to hold the increase in global average temperature to well below 2°C above pre-industrial levels and pursue efforts to limit the temperature increase to 1.5°C above pre-industrial levels. This will require considerable expansion of renewable power. At the same time, the Paris Agreement notes "the importance of ensuring the integrity of all ecosystems, including oceans, and the protection of biodiversity [...] when taking action to address climate change".
Convention on Biological Diversity	In the Kunming-Montreal Global Biodiversity Framework, targets 14-23 are on "Tools and solutions for implementation and mainstreaming". Target 14, for example, is to "Ensure the full integration of biodiversity and its multiple values into policies, regulations, planning and development processes [...] strategic environmental assessments, environmental impact assessments [...] within and across all levels of government and across all sectors". In the 2011-2020 Strategic Plan for Biodiversity, Strategic Goal A covering Aichi Targets 1-4 was: "Address the underlying causes of biodiversity loss by mainstreaming biodiversity across government and society". Decision XIV/3 focuses on mainstreaming biodiversity into the energy sector and infrastructure (in addition to mining, manufacturing and processing sectors).
Convention on the Conservation of Migratory Species of Wild Animals (CMS)	Resolution 7.4 Electrocution of Migratory Species, Resolution 7.5 Wind Turbines and Migratory Species, Resolution 10.11 Power Lines and Migratory Species, Resolution 11.27 Renewable power and Migratory Species, and Resolution 12.21 Climate Change and Migratory Species.
Ramsar Convention on Wetlands	Decision XI.10 Wetlands and energy issues.
Other relevant international agreements	
2030 Agenda for Sustainable Development and the SDGs	SDG 7. Ensure access to affordable, reliable, sustainable and modern energy for all, which includes a target (7.2) to increase substantially the share of renewable power in the final energy consumption. SDG 9. Build resilient infrastructure, promote inclusive and sustainable industrialization and foster innovation. SDG 13. Take urgent action to combat climate change and its impacts. SDG 14. Conserve and sustainably use the oceans, seas and marine resources for sustainable development. SDG 15. Protect, restore and promote sustainable use of terrestrial ecosystems, sustainably manage forests, combat desertification, and halt and reverse land degradation and halt biodiversity.
OECD Legal Instruments	OECD Council Recommendation on the Governance of Infrastructure. OECD Council Recommendation concerning the Reduction of Environmental Impacts from Energy Production and Use.
Group of Twenty (G20)	2019 Osaka Leaders' Declaration of the Group of Twenty (G20) endorsed a set of six voluntary Principles for Quality Infrastructure Investment. Principle 3 Integrating Environmental Considerations in Infrastructure Investments addresses biodiversity.
Group of Seven (G7)	In 2016 G7 Leaders adopted the Ise-Shima Principles for Promoting Quality Infrastructure Investment. Principle 3 is on Addressing social and environmental impacts of infrastructure.

Note: See Annex A for more details.
Source: Author.

References

Allan, J., O. Venter and J. Watson (2017), "Temporally inter-comparable maps of terrestrial wilderness and the Last of the Wild", *Scientific Data*, Vol. 4/1, https://doi.org/10.1038/sdata.2017.187. [58]

Balvanera, P. et al. (2019), *IPBES Global Assessment Chapter 2.1 Status and Trends –Drivers of Change*, https://doi.org/10.5281/ZENODO.5517423. [4]

Barrett, D. (2015), *Wildlife induced outages and protection of overhead lines and substations*, INMR Congress, http://encron.hr/wp-content/uploads/2016/07/EPP-2481-9_15-15-Prekidi-uzrokovani-%C5%BEivotinjama-i-za%C5%A1tita-opreme-%C4%8Dlanak-ENG.pdf. [55]

Butt, N. et al. (2013), "Biodiversity Risks from Fossil Fuel Extraction", *Science*, Vol. 342/6157, pp. 425-426, https://doi.org/10.1126/science.1237261. [17]

CBD (2022), *Kunming-Montreal Global Biodiversity Framework*, https://www.cbd.int/doc/decisions/cop-15/cop-15-dec-04-en.pdf. [5]

CBD (1992), *Article 2 of the Convention on Biological Diversity*, https://www.cbd.int/convention/articles/?a=cbd-02. [56]

Dasgupta, P. (2021), *The Economics of Biodiversity: The Dasgupta Review*. [2]

Dashiell, S., M. Buckley and D. Mulvaney (2019), *Green Light Study: Economic and Conservation Benefits of Low-impact Solar Siting in California*. [50]

Diaz, S. et al. (2019), *Summary for policymakers of the global assessment report on biodiversity and ecosystem services of the Intergovernmental Science-Policy Platform on Biodiversity and Ecosystem Services*, IPBES, Bonn, https://www.ipbes.net/system/tdf/ipbes_7_10_add-1-_advance_0.pdf?file=1&type=node&id=35245 (accessed on 6 September 2019). [41]

Duffy, E. (ed.) (2013), "Biodiversity and ecosystem stability: a synthesis of underlying mechanisms", *Ecology Letters*, Vol. 16, pp. 106-115, https://doi.org/10.1111/ele.12073. [45]

Energy Transitions Commission (2023), *Streamlining planning and permitting to accelerate wind and solar deployment*, https://www.energy-transitions.org/wp-content/uploads/2023/01/Barriers_PlanningAndPermitting_vFinal.pdf. [52]

European Commission (2020), *Commission notice Guidance document on wind energy developments and EU nature legislation Commission notice Guidance document on wind energy developments and EU nature legislation Guidance document on wind energy developments and EU Nature Legislation*, European Commission (C2020) 7730 final, Brussels, https://ec.europa.eu/environment/nature/natura2000/management/docs/wind_farms_en.pdf (accessed on 3 May 2022). [29]

Gasparatos, A. et al. (2017), "Renewable energy and biodiversity: Implications for transitioning to a Green Economy", *Renewable and Sustainable Energy Reviews*, Vol. 70, pp. 161-184, https://doi.org/10.1016/j.rser.2016.08.030. [23]

Giam, X., J. Olden and D. Simberloff (2018), "Impact of coal mining on stream biodiversity in the US and its regulatory implications", *Nature Sustainability*, Vol. 1/4, pp. 176-183, https://doi.org/10.1038/s41893-018-0048-6. [19]

Griscom, B. et al. (2017), “Natural climate solutions”, *Proceedings of the National Academy of Sciences*, Vol. 114/44, pp. 11645-11650, https://doi.org/10.1073/pnas.1710465114. [42]

Harfoot, M. et al. (2018), “Present and future biodiversity risks from fossil fuel exploitation”, *Conservation Letters*, Vol. 11/4, p. e12448, https://doi.org/10.1111/conl.12448. [18]

Harris, R. et al. (2018), *Biological responses to the press and pulse of climate trends and extreme events*, Nature Publishing Group, https://doi.org/10.1038/s41558-018-0187-9. [13]

Holland, R. et al. (2019), “The influence of the global electric power system on terrestrial biodiversity”, *Proceedings of the National Academy of Sciences*, Vol. 116/51, pp. 26078-26084, https://doi.org/10.1073/pnas.1909269116. [20]

Hund, K. et al. (2020), *Minerals for Climate Action: The Mineral Intensity of the Clean Energy Transition*, International Bank for Reconstruction and Development//The World Bank, https://pubdocs.worldbank.org/en/961711588875536384/Minerals-for-Climate-Action-The-Mineral-Intensity-of-the-Clean-Energy-Transition.pdf (accessed on 14 October 2021). [37]

IEA (2023), *World Energy Outlook 2023*, https://iea.blob.core.windows.net/assets/42b23c45-78bc-4482-b0f9-eb826ae2da3d/WorldEnergyOutlook2023.pdf. [9]

IEA (2022), *2022 IEA Ministerial Communiqué*, https://www.iea.org/news/2022-iea-ministerial-communique. [57]

IEA (2021), *The Role of Critical Minerals in Clean Energy Transitions*, IEA, https://www.iea.org/reports/the-role-of-critical-minerals-in-clean-energy-transitions. [36]

IEA (2021), *The Role of Critical Minerals in Clean Energy Transitions*, International Energy Agency, https://www.iea.org/reports/the-role-of-critical-minerals-in-clean-energy-transitions. [31]

IEA (2021), *World Energy Outlook 2021*, OECD Publishing, Paris, https://doi.org/10.1787/14fcb638-en. [7]

IPBES-IPCC (2021), *Scientific outcome IPBES-IPCC Co-sponsored Workshop on Biodiversity and Climate Change*. [12]

IPCC (2022), *Climate Change 2022: Mitigation of Climate Change. Working Group III contribution to the Sixth Assessment Report of the Intergovernmental Panel on Climate Change*. [6]

IPCC (2018), *Global Warming of 1.5°C*. [11]

Kiesecker, J. et al. (2019), “Hitting the Target but Missing the Mark: Unintended Environmental Consequences of the Paris Climate Agreement”, *Frontiers in Environmental Science*, Vol. 0, p. 151, https://doi.org/10.3389/FENVS.2019.00151. [47]

Luderer, G. et al. (2019), “Environmental co-benefits and adverse side-effects of alternative power sector decarbonization strategies”, *Nature Communications 2019 10:1*, Vol. 10/1, pp. 1-13, https://doi.org/10.1038/s41467-019-13067-8. [30]

McManamay, R., C. Vernon and H. Jager (2021), “Global Biodiversity Implications of Alternative Electrification Strategies Under the Shared Socioeconomic Pathways”, *Biological Conservation*, Vol. 260, p. 109234, https://doi.org/10.1016/J.BIOCON.2021.109234. [21]

Murphy-Mariscal, M., S. Grodsky and R. Hernandez (2018), "Solar Energy Development and the Biosphere", *A Comprehensive Guide to Solar Energy Systems*, pp. 391-405, https://doi.org/10.1016/B978-0-12-811479-7.00020-8. [24]

Narayan, S. et al. (2019), *Valuing the Flood Risk Reduction Benefits of Florida's Mangroves*, http://www.conservationgateway.org/SiteAssets/Pages/floridamangroves/Mangrove_Report_digital_FINAL.pdf. [44]

Nunez, S. et al. (2019), "Assessing the impacts of climate change on biodiversity: is below 2 °C enough?", *Climatic Change*, Vol. 154/3-4, pp. 351-365, https://doi.org/10.1007/s10584-019-02420-x. [15]

OECD (2021), "Biodiversity, natural capital and the economy: A policy guide for finance, economic and environment ministers", *OECD Environment Policy Papers*, No. 26, OECD Publishing, Paris, https://doi.org/10.1787/1a1ae114-en. [3]

OECD (2020), "Nature-based solutions for adapting to water-related climate risks", *OECD Environment Policy Papers*, No. 21, OECD Publishing, Paris, https://doi.org/10.1787/2257873d-en. [43]

OECD (2019), *Accelerating Climate Action: Refocusing Policies through a Well-being Lens*, OECD Publishing, Paris, https://doi.org/10.1787/2f4c8c9a-en. [8]

OECD (2019), *Global Material Resources Outlook to 2060: Economic Drivers and Environmental Consequences*, OECD Publishing, Paris, https://doi.org/10.1787/9789264307452-en. [35]

OECD Ministerial Council (2020), *2020 Ministerial Council Statement: Strong, resilient, inclusive and sustainable recovery from COVID-19.* [1]

Oliver, T. et al. (2015), "Declining resilience of ecosystem functions under biodiversity loss", *Nature Communications*, Vol. 6/1, https://doi.org/10.1038/ncomms10122. [46]

Rehbein, J. et al. (2020), "Renewable energy development threatens many globally important biodiversity areas", *Global Change Biology*, Vol. 26/5, pp. 3040-3051, https://doi.org/10.1111/gcb.15067. [33]

Saikia, L. (ed.) (2022), "Land-use intensity of electricity production and tomorrow's energy landscape", *PLOS ONE*, Vol. 17/7, p. e0270155, https://doi.org/10.1371/journal.pone.0270155. [25]

Santangeli, A. et al. (2015), "Global change synergies and trade-offs between renewable energy and biodiversity", *GCB Bioenergy*, Vol. 8/5, pp. 941-951, https://doi.org/10.1111/gcbb.12299. [32]

Smith, P. et al. (2018), "Impacts on terrestrial biodiversity of moving from a 2°C to a 1.5°C target", *Philosophical Transactions of the Royal Society A: Mathematical, Physical and Engineering Sciences*, Vol. 376/2119, p. 20160456, https://doi.org/10.1098/rsta.2016.0456. [14]

Sonter, L. et al. (2020), "Renewable energy production will exacerbate mining threats to biodiversity", *Nature Communications 2020 11:1*, Vol. 11/1, pp. 1-6, https://doi.org/10.1038/s41467-020-17928-5. [38]

Sovacool, B. et al. (2020), "Sustainable minerals and metals for a low-carbon future", *Science*, Vol. 367/6473, pp. 30-33, https://doi.org/10.1126/SCIENCE.AAZ6003. [34]

Susskind, L. et al. (2022), "Sources of opposition to renewable energy projects in the United States", *Energy Policy*, Vol. 165, p. 112922, https://doi.org/10.1016/j.enpol.2022.112922. [53]

Tegen, S. et al. (2016), *An Initial Evaluation of Siting Considerations on Current and Future Wind Deployment*, National Renewable Energy Laboratory, https://www.nrel.gov/docs/fy16osti/61750.pdf (accessed on 2 November 2021). [51]

Trainor, A., R. McDonald and J. Fargione (2016), "Energy sprawl is the largest driver of land use change in United States", *PLoS ONE*, Vol. 11/9, https://doi.org/10.1371/JOURNAL.PONE.0162269. [28]

UN (2023), *The Sustainable Development Goals Report 2023: Special Edition Towards a Rescue Plan for People and Planet*, https://unstats.un.org/sdgs/report/2023/The-Sustainable-Development-Goals-Report-2023.pdf. [10]

UNECE (2021), *Carbon Neutrality in the UNECE Region: Integrated Life-cycle Assessment of Electricity Sources*, https://unece.org/sites/default/files/2022-04/LCA_3_FINAL%20March%202022.pdf. [22]

van de Ven, D. et al. (2021), "The potential land requirements and related land use change emissions of solar energy", *Scientific Reports 2021 11:1*, Vol. 11/1, pp. 1-12, https://doi.org/10.1038/s41598-021-82042-5. [27]

van de Ven, D. et al. (2021), "The potential land requirements and related land use change emissions of solar energy", *Scientific Reports*, Vol. 11/1, https://doi.org/10.1038/s41598-021-82042-5. [48]

van Zalk, J. and P. Behrens (2018), "The spatial extent of renewable and non-renewable power generation: A review and meta-analysis of power densities and their application in the U.S.", *Energy Policy*, Vol. 123, pp. 83-91, https://doi.org/10.1016/J.ENPOL.2018.08.023. [26]

Warren, R. et al. (2018), "The projected effect on insects, vertebrates, and plants of limiting global warming to 1.5°C rather than 2°C", *Science*, Vol. 360/6390, pp. 791-795, https://doi.org/10.1126/science.aar3646. [16]

WEF (2020), *Nature Risk Rising: Why the Crisis Engulfing Nature Matters for Business and the Economy*, World Economic Forum, https://www3.weforum.org/docs/WEF_New_Nature_Economy_Report_2020.pdf. [39]

Wind Europe (2017), *Mainstreaming energy and climate policies into nature conservation: The Role of Wind Energy in Wildlife Conservation*, https://windeurope.org/wp-content/uploads/files/policy/topics/sustainability/WindEurope-Paper-on-the-role-of-wind-energy-in-wildlife-conservation.pdf (accessed on 2 November 2021). [54]

Wu, G. et al. (2023), "Minimizing habitat conflicts in meeting net-zero energy targets in the western United States", *Proceedings of the National Academy of Sciences*, Vol. 120/4, https://doi.org/10.1073/pnas.2204098120. [49]

WWF (2022), *Living Planet Index*, https://livingplanet.panda.org/#:~:text=The%20Living%20Planet%20Report%202022,are%20to%20reverse%20nature%20loss. [40]

Notes

[1] Biological diversity or biodiversity refers to "means the variability among living organisms from all sources including, inter alia, terrestrial, marine and other aquatic ecosystems and the ecological complexes of which they are part; this includes diversity within species, between species and of ecosystems" (CBD, 1992[56]).

[2] The extent of Wilderness Areas is based on the 'Last of the Wild' map, which identifies the most ecologically intact places on Earth. To produce the map, (Allan, Venter and Watson, 2017[58]) identified the 10% (by area) of each of the Earth's Biogeographic Realms with the lowest Human Footprint. The Human Footprint is a globally standardized map of cumulative human pressure on the natural environment. From this, all contiguous areas >10 000 km^2 were selected, in Biorealms that did not have 10 contiguous blocks >10 000 km^2, the next largest patch was consecutively selected until there were 10 per Biorealm. The final map contains 834 contiguous wilderness areas.

[3] Mining areas were mapped using a 50-cell radius around 62 381 pre-operational, operational, and closed mining properties. Each cell represents 1 km. The 50-cell radius is based on the assumption that impacts can extend 50 km from mine sites.

[4] At the IEA Ministerial Meeting in March 2022, IEA Member Countries voted to endorse and deepen the IEA's work on critical minerals, including by "investigating [...] different methods of ensuring the availability, security and responsible sourcing of energy-specific critical minerals" (IEA, 2022[57]).

[5] The 109 countries represent 83% of global terrestrial lands, and 92% of global GHG emissions.

3 Biodiversity impacts of solar power, wind power and power lines

This chapter examines the evidence of biodiversity impacts from solar photovoltaics, concentrated solar power, onshore wind, offshore wind and power lines. It first reviews the evidence of biodiversity impacts resulting from the construction, operation and decommissioning of renewable power infrastructure. It then discusses the importance of considering upstream impacts from the extraction and processing of the minerals required for renewable power infrastructure.

Understanding and mitigating the impacts of renewable energy infrastructure on biodiversity is fundamental to ensuring electricity expansion is not just low-emissions, but also biodiversity-aligned. Renewable energy and power lines can impact biodiversity at various stages of the infrastructure life cycle, from the mining and processing of minerals required for infrastructure components through to the construction, operation, maintenance and eventual decommissioning or repowering of energy facilities (Figure 3.1).

These impacts on biodiversity can be positive or negative, direct or indirect. Direct impacts refer here to those impacts resulting directly from renewable power and grid infrastructure projects, or the mining of resources necessary for their construction. Examples include habitat loss and fragmentation from mining or the construction of a renewable power facility, direct species mortality from collision with infrastructure, habitat creation and changes to ecosystem service provision or access (e.g., recreational or aesthetic values of a landscape; food provisioning services). Direct impacts tend to be easier to account for in decision-making than indirect and cumulative impacts. This is because direct impacts tend to be predictable, occur during the lifetime of an infrastructure project and usually occur at or close to the site where renewable power infrastructure is deployed (Bennun et al., 2021[1]).

Indirect impacts are the by-products or induced effects of renewable power deployment. A positive indirect impact on biodiversity may arise where renewable power expansion displaces fossil fuels, thereby helping to mitigate climate change. Another example of a positive indirect impact is where improving rural access to renewably sourced electricity decreases reliance on forests for fuel, thereby reducing deforestation (Tanner and Johnston, 2017[2]). Adverse indirect impacts may occur when the development of access roads for mining or renewable power projects facilitate (illegal) logging and other natural resource extraction, or economic opportunities surrounding mining and renewable power development lead to in-migration and subsequent increases in pressure on local biodiversity. Indirect impacts may occur far from the site where renewable power is deployed (e.g., from displacement of agriculture or other human activities). They are influenced by multiple factors external to a specific project or programme and are less predictable than direct impacts (Bennun et al., 2021[1]).

The various direct and indirect impacts can combine to cause cumulative impacts. Cumulative impacts include the combined impacts of a single renewable power facility (e.g., habitat fragmentation and direct mortality), the combined impacts of multiple projects, either from the same sector or multiple sectors across an ecosystem, landscape or migratory route, and the combined effects of pressures over time. Cumulative impacts may be additive (i.e. the impact is equal to the sum of individual impacts), synergistic (i.e. the cumulative impact is greater than the sum of the individual impacts), or antagonistic (i.e. the cumulative impact is less than the sum of its individual impacts) (Whitehead, Kujala and Wintle, 2017[3]; IFC, 2013[4]; Goodale and Milman, 2019[5]).

This chapter examines the direct, indirect and cumulative impacts of solar energy photovoltaics (PV), concentrated solar power (CSP), wind energy (onshore and offshore) and electricity networks (transmission and distribution lines). It first synthesises the evidence for biodiversity impacts arising during their construction, operation, maintenance and decommissioning. It then discusses the importance of considering the upstream impacts of renewable power infrastructure that result from the extraction and processing of minerals required for infrastructure components.

Figure 3.1. Overview of life stages of renewable power infrastructure

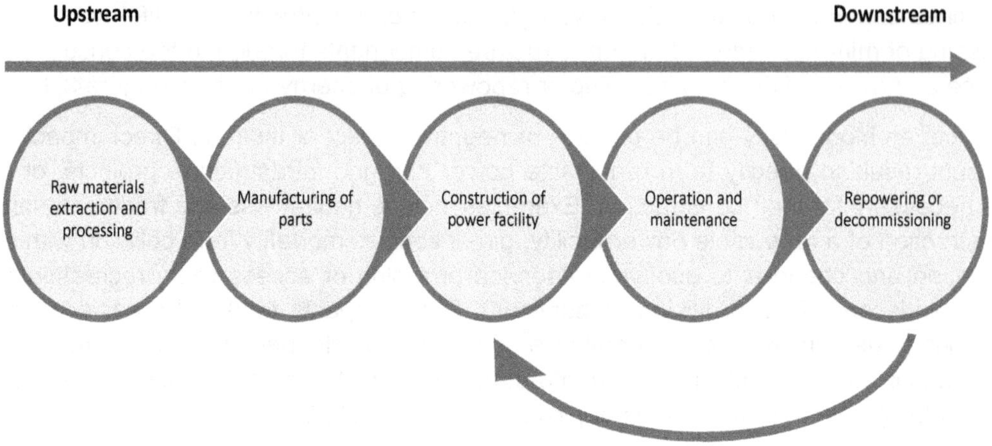

Source: Author.

3.1. Synthesis of evidence for biodiversity impacts: construction to decommissioning

Renewable power infrastructure can affect biodiversity in various ways during its construction, operation and maintenance, and decommissioning or repowering (Table 3.1). The type and magnitude of impacts depend on the renewable power technology. For example, bird and bat collision are a particular concern for power lines and wind turbines, whereas habitat loss is a greater concern for solar energy. Impacts also depend heavily on site-specific variables (i.e., location), and the practices employed to design, construct, operate, maintain and decommission renewable power infrastructure. As discussed throughout this report, opportunities exist to scale up wind and solar power without compromising biodiversity goals provided biodiversity is effectively mainstreamed throughout planning, policy and project processes.

The body of evidence for renewable power impacts on biodiversity has developed considerably in recent years, and is helping to inform efforts by governments, energy developers and investors to mitigate negative biodiversity impacts. However, data and knowledge of biodiversity impacts are uneven across technologies, taxa and geographies. Specifically, the evidence base for onshore wind is more established than for offshore wind and solar energy. Whereas impacts on avian species (e.g., collision mortality) have been extensively studied (particularly for wind), few studies examine impacts on other taxa. Most of the evidence for biodiversity impacts comes from parts of Europe and North America; the evidence base is less developed in regions where significant expansion of renewable power infrastructure is expected (e.g., Africa, Latin America, Southeast Asia).

For solar PV and CSP, onshore and offshore wind and electricity networks, key knowledge gaps remain concerning population-level and cumulative impacts (e.g., for migratory species), knock-on effects on ecological communities, ecosystems and ecosystem services, indirect impacts and the effectiveness of mitigation measures. Focused research efforts, long-term systematic monitoring and standardised data collection will be important for addressing knowledge gaps and understanding how impacts may scale with the global expansion in renewable power (see 5.1.4).

Table 3.1. Summary of potential biodiversity impacts of solar power, wind power and power lines

Non-exhaustive list of examples from literature

Potential biodiversity impacts from renewable power infrastructure	Solar energy	Wind energy	Power lines (solar and wind)
Direct wildlife mortality and morbidity	• Avian collision with panels or mirrors • Burning of birds and insects (CSP) • Potential bat and insect collision with fans of air-cooled condensers (CSP) • Drowning or poisoning of birds, reptiles and mammals in evaporation ponds (CSP)	• Avian and bat collision with turbines • Potential (secondary) entanglement of whales, turtles, diving sea birds and some fish with cables, anchors and mooring lines (floating offshore wind)	• Avian collision with power lines • Electrocution of birds, mammals, reptiles
Habitat loss and degradation	• Vegetation clearance and management where solar panels/mirrors are installed • Change in surface-water flows • Impacts on aquatic biodiversity in water-scarce areas due to water consumption for cooling (CSP)	• Vegetation clearance or disturbance for turbine foundations, access roads etc. e.g., forest clearing (onshore) • Loss/degradation of benthic habitat from installation of anchors, foundations and cables (offshore)	• Vegetation clearance and management of RoW under over-ground cables • Vegetation clearance and earth removal for underground cables
Habitat fragmentation and barrier effects	• Physical barrier to non-volant animals • Possible edge effects	• Barrier effects for birds and bats, particularly migratory species	• Barrier effect e.g., for birds and arboreal mammals • Edge effects
Habitat alteration / creation (potentially positive or negative impacts on biodiversity depending on context)	• Microclimatic changes created by solar panels (e.g., shading) can change species composition, richness and diversity • Panels may provide nesting sites or shelter for some bird species, arthropods and plants	• Wind turbine foundations can create a “reef effect” (offshore)	• Pylons and RoWs used for nesting and foraging by some bird species • RoWs provide corridor for some species to move across the landscape • RoWs support different ecological communities
Behavioural changes, species displacement and physiological changes	• Avoidance of solar facilities during construction or operation • Attraction to solar facilities (e.g., aquatic insect attraction to polarised light; waterbirds mistaking panels for water)	• Avoidance of wind facilities during construction or operation • Attraction to wind facilities • Physiological stress from operation of facilities	• Avoidance of power lines by birds and some mammals (effective habitat loss) • Electromagnetic field effects on behaviour (attraction/avoidance), physiology and navigation • (Temporary) displacement from noise during construction
Potential impacts from invasive alien species	• IAS could be introduced during construction and colonise more easily due to practices of vegetation clearance, mowing etc.	• IAS could be introduced during construction (e.g., by shipping vessels or trucks) and colonise and disperse more easily in disturbed areas (e.g. across access roads of onshore facilities and foundations of offshore facilities)	• Colonisation and dispersal of IAS along RoW and under pylons
Ecosystem service impacts (potentially positive or negative impacts on biodiversity depending on context)	• Impacts on carbon sequestration, nutrient and hydrological cycles, pollination services, habitat	• Impacts on aesthetic values, affecting tourism • Potential carbon loss e.g., where forests converted for	

Potential biodiversity impacts from renewable power infrastructure	Solar energy	Wind energy	Power lines (solar and wind)
	provision, water and food supply, and cultural values	wind energy • Food provisioning affected e.g., from exclusion of fisheries in offshore facilities	
Indirect impacts	• Impacts could arise if agriculture displaced • Displacement of more greenhouse gas intensive energy sources, thereby helping mitigate climate-related impacts • Reduced pressure on ecosystems in developing countries by facilitating alternative livelihoods (e.g., stopping charcoal production, reducing slash and burn agriculture)	• Impacts could arise if fisheries displaced (offshore) • Displacement of more greenhouse gas intensive energy sources, thereby helping mitigate climate change • Reduced pressure on ecosystems in developing countries by facilitating alternative livelihoods (e.g., stopping charcoal production, reducing slash and burn agriculture)	• Increased fire risk • Increased deforestation and hunting by facilitating access • Displacement of more greenhouse gas intensive energy sources, thereby helping mitigate climate change • Reduced pressure on ecosystems in developing countries by facilitating alternative livelihoods (e.g., stopping charcoal production, reducing slash and burn agriculture)
Cumulative and population-level impacts	• Population-level effects likely for some sensitive species (e.g., grassland birds in US) owing to cumulative impacts of multiple energy developments • Some ecosystems (e.g., desert and xeric shrubland) may experience significant cumulative impacts owing to concentration of solar energy facilities	• Cumulative impacts on populations of sensitive bird and bat species due to collision • Cumulative impacts on marine species and ecosystems (offshore)	• Cumulative impacts on populations of sensitive species due to collision and electrocution • Cumulative habitat loss and fragmentation (e.g., in Amazon rainforest)

Note: This table provides examples of potential impacts based on empirical evidence and inference. Mitigation measures can avoid or significantly reduce the severity of these impacts. CSP = concentrated solar power. IAS = invasive alien species. RoW = right of way.
Source: Author based on numerous references (see text below for specific references).

3.1.1. Solar energy

The two main types of solar energy technologies are PV and CSP. Solar PV is much more geographically widespread than CSP, and accounts for most solar capacity and projected growth in solar capacity. Solar PV cells are assembled in solar panels (modules) to convert sunlight into direct electric current, using the photoelectric effect. Multiple solar panels can be connected to form solar arrays. Solar modules can be mounted on the ground, on existing infrastructure, such as individual residential and commercial buildings, or on purpose-built floating structures.

Solar modules may be fixed-mount or automated tracking systems that follow the sun's path. Solar tracking systems optimise electricity production but tend to have higher capital costs. While they generally require more land surface than fixed-mount systems to avoid shading, the higher output of solar tracking systems means that fewer panels are need for the same energy outputs (Campbell et al., 2009[6]; Dunlop, 2010[7]).

PV facilities vary in size from small schemes of <1 megawatt (MW), which provide energy to a single consumer or a small group of consumers, to large utility-scale facilities with a capacity of more than 1 MW and as much as 2 gigawatts (GW) (e.g. Huanghe Hydropower Hainan Solar Park China, which spans 564 hectares) (Murray, 2021[8]; Hernandez et al., 2014[9]). Small schemes are often incorporated into existing infrastructure, such as rooftop solar schemes (see 4.3.1).

While the majority of PV capacity is land-based, the installed capacity of floating solar photovoltaic facilities is increasing. Floating PV facilities are typically installed on freshwater lakes, fabricated reservoirs and canals but may also be installed in various marine ecosystems. For example, marine solar PV demonstration projects have been established in the Persian Gulf (coastal waters of Dubai in the United Arab Emirates), in the North Sea (The Netherlands), deep fjords (Norway), and shallow tropical lagoons (Maldives) (Hooper, Armstrong and Vlaswinkel, 2021[10]). Floating solar could help reduce land-use pressure while increasing solar panel efficiency due to the cooling effect of water on PV modules, but it too can generate risks to biodiversity that need to be assessed and managed (Armstrong et al., 2020[11]; Dörenkämper et al., 2021[12]; Sacramento et al., 2015[13]; Choi, 2014[14]).

The other solar energy technology, CSP, uses mirrors to concentrate the sun's energy onto a receiver that converts it to heat. The heat can then be used to create steam, which drives a turbine to generate electricity. Several CSP technologies exist, including arrays of mirrors (heliostats) that track the sun and concentrate light on a fixed centralised receiver (solar power tower), arrays of linear mirrors (Fresnel reflectors) laid flat on the ground that focus light on liquid-filled pipes, parabolic solar troughs that focus light on a receiver running along their focal point, and parabolic dish systems comprising standalone parabolic reflectors that concentrate light on a receiver at the focal point (SolarPACES, 2018[15]; Brunel, 2021[16]). Unlike PV facilities, CSP plants can store thermal energy that can then be converted into electricity and dispatched in response to demand, even during the night or cloudy periods. CSP facilities can provide baseload power around the clock and be ramped up quickly in response to grid requirements (IBRD, 2020[17]).

Utility-scale PV and CSP facilities have similar infrastructure requirements. Requirements include module mounting infrastructure and associated solar tracking systems, solar panels or mirrors (reflectors), electricity infrastructure such as cabling, inverters, transformers, and on-site sub-station and transmission lines to connect to the power grid, access roads and security perimeter fences. In addition to these common components, CSP facilities require a concentrating solar collector such as solar power towers (Bennun et al., 2021[1]). Floating PV facilities need a floating support structure for solar arrays and a mooring and anchoring system (Exley et al., 2021[18]).

Ground-mounted solar facilities require large areas of land to accommodate infrastructure such as solar panels (PV) or mirrors and towers (CSP). Lovering et al. (2022[19]) find that on average ground-mounted utility scale CSP require 2 000 hectares per terawatt hour per year (TWh/y) and utility scale PV requires 2 100 hectares per TWh/y. Owing to their relatively large land requirements, ground-mounted utility scale facilities are sometimes sited far from end-users. Their remoteness increases the requirement for additional infrastructure such as power line corridors, roads and substations, which can increase the overall physical footprint of solar energy. Increasing improvements in solar energy efficiency could help to reduce the amount of area required per unit of energy produced. PV technologies that are integrated into the built environment are among the most land-use efficient source of renewable power (Fthenakis and Kim, 2009[20]).

Potential impacts of solar power facilities on biodiversity include direct wildlife morbidity and mortality, habitat loss and degradation, habitat fragmentation and barrier effects, habitat alteration or creation, behavioural changes, physiological changes and displacement, IAS impacts, ecosystem service impacts, indirect impacts and cumulative population-level impacts. Peer-reviewed empirical evidence for these impacts is scarce, however, and reviews of impacts have often inferred impacts. While ground-mounted solar energy development can negatively impact biodiversity, studies suggest that in certain contexts (e.g., when installed in degraded lands and with biodiversity-specific management practices), solar facilities can be deployed with minimal or potentially even positive impacts on biodiversity. The impacts of solar energy on biodiversity are discussed in detail below.

Direct wildlife morbidity and mortality

The presence and operation of solar energy facilities can lead to direct wildlife morbidity and mortality. The nature and magnitude of impacts depend on the technology (e.g., PV or CSP), and a project's location, size and design (Walston et al., 2016[21]). The two primary causes of direct mortality in solar energy facilities are collision and burning. Drowning of species in solar evaporation ponds of CSP have also been documented (Jeal et al., 2019[22]). Avian mortality from collision with solar facility infrastructure has been documented for both PV and CSP sites of all technology types. In addition, several studies indicate that bats and insects may also collide with solar infrastructure, particularly fans of air-cooled condensers of CSP sites (Murphy-Mariscal, Grodsky and Hernandez, 2018[23]). Collision risk is likely to be higher when surfaces are oriented vertically and reflecting light (Bennun et al., 2021[1]).

Given the sparseness of data in peer-reviewed literature, generalisations of collision risk are limited for solar facilities. Unlike for wind energy and other energy infrastructure (e.g., power lines), it is unclear which bird species are at higher risk of collision. However, mortality from collision has been documented in a wide range of avian species. For example, Kosciuch et al. (2020[24]) documented avian mortality across nine taxonomic orders. In the grey literature, Kagan et al. (2014[25]) found that mortality at three solar facilities in California (a CSP tower facility (Ivanpah), CSP trough (Genesis) and PV (Desert Sunlight)), occurred in species of different body sizes (e.g. hummingbirds and pelicans), and with markedly different ecology (e.g. aerial feeders, aquatic feeders, ground feeders and raptors; nocturnal and diurnal species; and resident and non-resident species). Solar facilities may pose a risk to all birds flying over or using a solar energy facility, but the level of risk likely depends on biological, topographical, meteorological and technical factors (Visser et al., 2019[26]).

The extent or significance of collision mortality for solar energy is poorly understood but is likely to be lower than mortality from fossil fuel facilities (Walston et al., 2016[21]), wind turbines, power lines and other human activities at the current capacity. Extrapolating findings from a three-month study at the largest solar PV facility in South Africa, Visser et al. (2019[26]) estimated a mortality rate of 4.5 birds/MW/year, although the cause of mortality could not be established. A study of carcasses across 10 PV plants in California and Nevada, US, spanning 13 site-years,[1] found collision was the primary determinable[2] cause of mortality, but was unable to identify the cause of mortality for 61% of carcasses. Estimates of avian mortality within the solar field (i.e. excluding fences and power lines) of these facilities ranged from 0.08 birds/MW/year (0.03 birds/hectare/year) to 9.26 birds/MW/year (5.17 birds/hectare/year), with a mean of 2.49 birds/MW/year (1.09 birds/hectare/year) (Kosciuch et al., 2020[24]).

Walston et al (2016[21]) provided a significantly higher estimate of avian mortality from solar facilities in the US, of 9.9 birds per MW per year. Three factors partly explain the higher estimate (Kosciuch et al., 2020[24]). First, the study covered different solar technologies, one PV facility (California Valley Solar Ranch) and two CSP tower facilities (California Solar One and Ivanpah). CSP tower facilities pose not only a collision risk, but also a risk of burning from solar flux (discussed below) and were found to have a mortality rate 7-21 times higher than at the PV site. Second, the study included all infrastructure monitored (e.g., power lines and fences). Third, the PV facility included had significantly higher annual mortality rate than the other 12 site-years included in the study by Kosciuch et al. (2020[24]). While these three studies provide an indication of the magnitude of collision mortality risk, it is important to note that their sample size is small and geographic scope limited. Mortality risk is not well understood in different habitat contexts, so attempts to extrapolate mortality rates to other projects may be misleading (Kosciuch et al., 2020[24]).

Another cause of morbidity and mortality, which is limited to CSP facilities, is burning. Birds and insects can be burned when they cross the concentrated solar light reflected to the central receiver in CSP facilities. When birds' flight feathers are singed from solar flux, their flight can be impaired, increasing the risk of collision with the ground or other objects and reducing their capacity to feed and avoid predators (Kagan et al., 2014[25]; Walston et al., 2016[27]). Each of these effects can lead to mortality. Avian morbidity and mortality from burning has been documented at CSP facilities in Israel, Spain and the US (Ho, 2016[28];

Kagan et al., 2014[25]). Fatal burning of insects such as dragonflies and butterflies has also been observed (Kagan et al., 2014[25]), but has not been addressed in peer-reviewed literature.

Evaporation ponds at solar facilities, used to store wastewater and concentrate chemicals before disposal, can also pose a risk to wildlife from drowning or poisoning. A four-month study of a CSP plan in South Africa, for example, identified 37 carcasses in evaporation ponds, including four species of birds, one species of reptile and seven species of mammals (Jeal et al., 2019[22]).

Given the scarcity of peer-reviewed data and limited understanding of the mechanisms leading to wildlife morbidity and mortality at solar facilities, further empirical research would be beneficial. Such research could help to ensure solar facilities are located, constructed and operated in a way that is consistent with biodiversity objectives. Systematic, repeatable and standardised sampling protocols could help to build the evidence base while promoting accuracy, precision and comparability (Huso, Dietsch and Nicolai, 2016[29]; Visser et al., 2019[26]). In addition, behavioural studies could help improve understanding of species risk factors and identify mitigation strategies. Priority research areas relate to species' perception of solar facilities (attraction/deterrence), movement, habitat use and interspecific interactions to inform mitigation measures (Chock et al., 2020[30]).

Habitat loss and degradation

Habitat loss and degradation can occur at the construction, operation and decommissioning phases of solar energy facilities. The construction of ground-mounted solar PV and CSP facilities may involve vegetation removal and surface grading to facilitate installation, prevent shading of solar panels by vegetation or undulating land and reduce on-site risk of wildfire. During the operation phase, some solar facilities apply herbicides, cover the land with gravel and mow frequently to manage the vegetation around solar panels (Turney and Fthenakis, 2011[31]).

These construction and operation practices can drive habitat loss and degradation, resulting in species mortality or displacement, which in turn can lead to declines in species richness and density (Murphy-Mariscal, Grodsky and Hernandez, 2018[32]). In addition to removal of plant species, clearance and grading can increase soil erosion and reduce the amount of organic carbon and nitrogen, which in turn can affect primary production by plants and food availability for wildlife (Antonio Sánchez-Zapata et al., 2016[33]). In the French Mediterranean, Lambert et al. (2021[34]) surveyed soil temperature and moisture, CO_2 effluxes, and vegetation below and outside solar panels of three solar parks. Physical, chemical and general soil quality indexes were lower in a solar park than in semi-natural land cover types (pinewood and shrubland). Clearing and grading the soil surface during solar park construction resulted in a strong degradation of soil physical quality, especially of soil structure.

Whether habitat loss and degradation from a solar facility is significant depends on the intactness and ecological value of the habitat prior to construction (i.e., the baseline), the location of the solar facility and how it is constructed and operated. PV or CSP solar facility deployment is more likely to result in biodiversity loss if it occurs in relatively undeveloped areas. In the US, for example, the perennial plant cover and structure at a CSP facility in California was found to be less than surrounding areas of relatively undeveloped desert (Grodsky and Hernandez, 2020[35]).

In contrast, solar facilities installed in degraded lands and actively managed for biodiversity may have a positive impact on habitat. For example, a study of eleven solar farms in the UK found that, overall, the facilities supported a higher diversity and abundance of broad leaved plants, grasses, birds and invertebrates than the agricultural or brownfield land where they were sited (Montag, Parker and Clarkson, 2016[36]). The degree to which these eleven solar facilities were beneficial to biodiversity was highly dependent on the site management approach: solar farms with the highest wildlife value were seeded with diverse seed mix after construction, limited the use of herbicides, provided marginal habitat for wildlife and adopted biodiversity-minded livestock grazing or mowing regimes.

The deployment of PV and CSP solar facilities may also degrade habitats by altering water quality and quantity. Surface-water flows are sometimes deliberately modified at solar farms to reduce soil erosion around solar infrastructure. This may affect downstream aquatic ecosystems and habitats by changing the flow of organic matter, nutrients, minerals, and sediments. Dust suppressants and herbicides used to maximise exposure of panels to sun can increase run-off and affect the chemical composition of waterways (Cameron, Cohen and Morrison, 2012[37]; Grippo, Hayse and O'Connor, 2014[38]). Cooling water released from CSP facilities could affect the temperature of water bodies and contaminate them with hazardous chemicals, such as cooling system toxicants, antifreeze agents, heavy metals and rust inhibitors (Hernandez et al., 2014[9]). This could be harmful to freshwater species (Bennun et al., 2021[1]).

In addition, utility scale solar may lead to water withdrawal and consumption. CSP facilities with wet-cooling systems require vast quantities of water: around 3 500 litres/MWh of electricity generated, compared to 1 000 litres/MWh in natural gas-fired power plants (EC, 2019[39]). Such facilities could place stress on water resources and affect riparian habitats particularly in water-scarce areas, which tend to be where CSP are located (Lovich and Ennen, 2011[40]). Dry-cooling or hybrid-cooling systems significantly reduce CSP water consumption (Hernandez et al., 2014[9]), but can reduce the efficiency of solar plants. Radiative cooling technologies are emerging that may help reduce water use of wet-cooling systems without compromising efficiency (Aili et al., 2022[41]). Both PV and CSP may use water for cleaning panels and mirrors or for dust suppression, although cleaning requires relatively small amounts of water and emerging technologies and practices are increasing the efficiency of cleaning (Hernandez et al., 2014[9]; EC, 2019[42]).

Generally, water withdrawal and consumption from solar PV entails relatively low risks and is much lower than alternative energy systems. For example, solar PV and wind turbines, across their life cycle, consume about 0.1–14% and withdraw about 2–15% of the water typically used by thermo-electric power plants (coal or nuclear) to generate 1 MWh of electricity (Roehrkasten, Schaeuble and Helgenberger, 2015[43]).

Habitat fragmentation and barrier effects

At a landscape level, the deployment of solar energy facilities may fragment habitat and create barriers to species movement. Habitat fragmentation is when a continuous habitat is divided into isolated patches of remnant habitat because of conversion or disturbance (Wilson et al., 2015[44]). It results in both a smaller total amount of habitat area and changes to a habitat's spatial configuration (Berger-Tal and Saltz, 2019[45]). Habitat fragmentation has been linked to declines in species richness, edge effects, compromised ecosystem function, and isolation of populations and reduced genetic exchange (Haddad et al., 2015[46]) (Fahrig, 2003[47]).

While the impacts of infrastructure-induced habitat fragmentation are relatively well-documented (e.g., for linear transport infrastructure), there is a dearth of studies focussing on the effects of solar infrastructure. Several studies, however, suggest that habitat fragmentation is an important consideration for solar energy expansion. One study, for example, found that over 70% of PV and 90% of CSP utility-scale installations planned and under construction in California, US, were within 10 km of a protected area (Hernandez et al., 2015[48]). The authors warned that the facilities could increase edge effects and undermine the effectiveness of the protected areas as wildlife corridors.

Another study found that the deployment of solar facilities in Florida, US, may undermine efforts to save the endangered Florida panther (*Concolor coryi*) (Leskova, Frakes and Markwith, 2022[49]). The only wild breeding population of the Florida panther is restricted to <5% of its historic range in South Florida, and the area may be close to carrying capacity. Three viable populations within the historic range are needed for species recovery. A comparison of Florida panther habitat suitability and connectivity pre- and post-installation of 45 utility scale solar energy facilities found that nine facilities were located within major corridors connecting the only breeding population with other areas that could support populations of Florida panther, 26 facilities were located within other areas that could facilitate some dispersal of the panther.

Such studies underscore the importance of considering landscape-level connectivity in strategic planning and siting decisions, in addition to environmental impacts within facility boundaries (see Chapter 3). Careful siting and design of solar facilities can reduce the extent of habitat fragmentation. For example, in a large solar power facility in central Australia, a vegetation strip was left as a north-south corridor through flour blocks of arrays, dividing the 250-ha block in two. This strip has reportedly enabled the movement of wildlife through the facility (Guerin, 2017[50]). In the light of climate-induced changes to species distribution and corridors, it is prudent to consider both current and future distribution and dispersal patterns of species when siting infrastructure.

Solar facilities tend to be secured by fences, which may provide a physical barrier to dispersal for non-volant animals, in addition to posing a collision risk for volant animals. Dispersal barriers can affect migration patterns, feeding and gene flow (McInturff et al., 2020[51]). The impact of solar facility fences on wildlife has not been quantified, but literature on fences used for other purposes provide an indication of the potential impacts and mitigation measures (Buton, 2023[52]), such as permeable fences to facilitate dispersal of kit foxes *Vulpes macrotis* in San Joaquin Valley, California (Cypher et al., 2021[53]).

Habitat alteration or creation

Solar panels or mirrors may also influence soil and microclimate conditions by catching precipitation and atmospheric deposition, changing surface albedo, increasing ground shading and affecting wind speed (Hernandez et al., 2014[9]). Owing to the albedo effect, for example, night temperatures at a solar PV installation in a rural area were found to be 3-4 degrees Celsius (°C) higher than wildlands (Barron-Gafford et al., 2016[54]). Conversely, in city environments where albedo is lower, modelling for the cities of Los Angeles (Taha, 2013[55]) and Paris (Masson et al., 2014[56]) suggested that PV deployment could have a net cooling effect. Owing to the insulation effect due to shading and airflow, spring and summer soil temperatures at a CSP plant were 0.5-4 °C lower in summer and higher in the winter compared to control sites with no collectors (Wu et al., 2014[57]). Similarly, a study at a UK PV facility observed cooling in the summer of up to 5.2 °C, and drying under the PV arrays, compared with gaps between PV arrays and control areas (Armstrong, Ostle and Whitaker, 2016[58]).

The impact of solar facilities on microclimates depends not only on their location but also the technology. For example, solar PV fixed-mount and tracking systems effect microclimatic conditions differently. A study of the two technologies in Chile determined that fixed-mount solar modules provide shade where the temperature is cooler and humidity is higher throughout the day, while solar tracking systems create temporally varying shading conditions (Suuronen et al., 2017[59]).

Changes to soil and microclimate conditions may lead to changes in species composition, richness and diversity (Tanner, Moore and Pavlik, 2014[60]). For example, a study of a UK solar PV facility found that species diversity and plant biomass under PV arrays were lower owing in part to differences in soil and air temperature (Armstrong, Ostle and Whitaker, 2016[58]). In some contexts, the shadow effect of solar panels could be beneficial, for example, when used to preserve crops during heatwaves and drought (Barron-Gafford et al., 2016[54]) or by providing shade and shelter for some arthropods (Suuronen et al., 2017[59]) and avian species (Visser et al., 2019[26]). If vegetation is allowed to regrow between panels, solar facilities could also provide nesting opportunities (Visser et al., 2019[26]). Further evidence of the microclimatic changes associated with solar facilities in different contexts, and the resulting impact on species and ecosystem services, could help to optimise solar facility design for biodiversity.

Behavioural changes, physiological changes and species displacement

The construction and operation of solar energy facilities could elicit behavioural responses for certain species. Potential responses include avoidance of (or attraction to) noise, light and physical structures, and changes to feeding patterns, competition and reproduction. Such responses, which may be temporary or permanent, can affect energy expenditure and predation risk, potentially reducing fecundity and increasing mortality in wildlife (Murphy-Mariscal, Grodsky and Hernandez, 2018[61]). Behavioural changes induced by solar energy facilities could, therefore, negatively affect populations and ecological communities.

Few studies have examined behavioural changes and species displacement induced by solar facilities. However, evidence suggests that solar energy facilities can displace species both through habitat destruction and avoidance behaviour (effective habitat loss). For example, a study of a CSP facility in South Africa found that birds were much more abundant (141.9 birds/km) and species rich (51 species) in the surrounding rangeland than in the solar field (1.27 birds/km; 22 species) (Jeal et al., 2019[22]). Another study found bird species richness (and to a lesser extent density) at South Africa's largest solar PV facility to be lower than the boundary zone and adjacent untransformed land (Visser et al., 2019[26]).

In some circumstances, species may be moved deliberately from a solar facility site prior to construction to avoid negative impacts on biodiversity. While this may reduce the risk to some species, mitigation-driven species translocation is not always successful and may put physiological stress on wildlife (Germano et al., 2015[62]). For example, desert tortoises *(Gopherus agassizii)* translocated to mitigate the impacts of a solar development in Mojave Desert in California, US, experienced higher body temperatures and increased energy expenditure during first year following displacement, particularly in the first month. However, translocation did not appear to affect the growth and body condition of the tortoises (Brand et al., 2016[63]). More recently, it was reported that 30 out of 139 tortoises relocated for the Yellow Pine Solar Energy project in Nevada died within a few weeks of their relocation (Castillo, 2021[64]).

While some species avoid solar energy facilities, others may be attracted to them. For example, preliminary research and anecdotal evidence suggests that solar PV can be "ecological traps" for insects. Aquatic insects may be attracted to the polarised light reflected by PV panels, mistaking the panels for water surfaces. The increased concentration of insects may then attract insectivorous birds and bats to the panels (Kagan et al., 2014[25]). The attraction of insects to solar PV could have knock-on effects on ecological communities, particularly where solar facilities are close to water bodies (Horváth et al., 2009[65]; Horváth, 2010[66]).

Unexpected findings of stranded, injured and deceased water-associated birds (i.e. species that rely on water for foraging, reproduction, and/or roosting, such as herons and egrets) and water-obligate birds (i.e. species that cannot take flight from land, such as loons and grebes) at a PV facility in California led scientists to propose that some birds mistake solar arrays for water (i.e. the Lake Effect Hypothesis) (Kagan et al., 2014[25]) (Kosciuch et al., 2020[24]). Little empirical evidence exists to support or disprove the hypothesis. A recent study exploring the hypothesis concluded that some species of aquatic birds could be attracted to solar PV facilities in certain contexts, but that a PV solar facility is "unlikely to provide a signal of a lake to all aquatic habitat birds at all times" (Kosciuch et al., 2021[67]).

Impacts from invasive alien species

Solar energy facilities could facilitate the introduction or spread of invasive alien species (IAS) through two general pathways, the first direct and the second indirect. First, the construction and maintenance of solar facilities involves the movement of solar energy infrastructure components, equipment and people, each of which provides a potential vector for species. For example, soil on machinery could introduce IAS to the facility. Second, the degradation or loss of habitat during construction or operation could reduce ecosystem resilience and thereby facilitate the colonisation or spread of IAS. In California, for example, invasive grasses have been found to take hold after blading (removal of above-ground vegetation) (Grodsky and Hernandez, 2020[35]). Data and literature on the colonisation or spread of IAS resulting from solar energy development (and the impact this has on ecological communities) are scarce.

Ecosystem services impacts

Solar energy deployment may affect ecosystem service supply and access. Potentially affected services include supporting services such as soil formation and nutrient cycles, regulating services such as climate and hydrology, provisioning services such as water and food supply, and cultural services such as recreational activities, aesthetic and spiritual values (Antonio Sánchez-Zapata et al., 2016[33]).

Quantification of solar energy impacts on ecosystem services is all but absent in the primary literature. However, some studies shine a light on the potential tensions between solar energy deployment and ecosystem service provision (Murphy-Mariscal, Grodsky and Hernandez, 2018[32]; Exley et al., 2021[68]; van de Ven et al., 2021[69]; Bevk and Golobič, 2020[70]; Hastik et al., 2015[71]). For example, De Marco et al. (2014[72]) deemed 42 of the 82 permitting requests for new utility scale solar energy sites in Lecce, Italy (equating to 18 563 ha of land-cover change), to be in ecologically unsuitable areas, owing to the ecosystem service values they put at risk. The 42 sites included century-old olive groves notable for their high cultural value and areas that provide a relatively large contribution to carbon sequestration, relative to other land-cover types evaluated. A study of Ivanpah Solar Electric Generating System in the Mojave Desert, California, found that non-bee insect flower visitors were negatively affected by the facility. The authors warned that solar energy disruption of non-bee insect flowers visitor communities in deserts could lead to cascading effects on biodiversity, including globally threatened pollinator-dependent cacti (Grodsky, Campbell and Hernandez, 2021[73]).

The impact of solar energy on ecosystem services depends on a facility's location and how it is developed and operated. For example, floating photovoltaics may affect nine ecosystem services linked to eight SDGs, however, whether the impact is positive or negative likely depends on the water body type and design of the system (Exley et al., 2021[68]). A study of the Ivanpah CSP facility in the Mojave Desert, California, showed ecosystem service values from desert plants differ among solar energy development decisions (Grodsky and Hernandez, 2020[35]). Provisioning, regulating, habitat and cultural ecosystem service values are lower in bladed treatments than mowed or halo[3] treatments, and highest at control sites. Bladed treatments result in ecosystem "disservices", by facilitating the colonisation of invasive grasses.

In some contexts, through proper siting, design and management, solar facilities could enhance several ecosystem services, while helping to combat climate change and meet energy demands (Randle-Boggis et al., 2020[74]; Walston et al., 2021[75]). For example, compared to pre-solar agricultural land uses, solar facilities in Midwest United States that restore and manage native grassland can increase pollinator supply by 300%, carbon storage potential by 65%, sediment retention by more than 95% and water retention by 19% (Walston et al., 2021[75]).

Indirect impacts

In theory, land take for solar facilities could displace other land uses. In some countries, solar energy has been deployed in agricultural or forestry land. For example, about 28% of utility scale solar energy in California is in agriculture land (cropland and pasture), equivalent to about 150 km^2 (Hernandez et al., 2015[48]). In some cases, solar and agriculture production may be co-located (4.3.1), thereby reducing land-use pressure. However, in other cases, solar energy deployed in agricultural land may come at the expense of agricultural production. This may shift agricultural activities to other locations to meet growing demand, thereby contributing to land-use change or pollution far from the solar project site.

When solar energy facilities are in remote, previously inaccessible locations, the construction of service roads could induce further road construction and facilitate access for other human activities. This could result in increased pressure on natural resources in the area, for example though legal and illegal logging, harvesting and hunting, and pollution. Increased movement of people facilitated by new transport routes could also facilitate the spread of invasive alien species. While such indirect impacts of road construction have been examined, particularly for tropical forest systems (Laurance, Goosem and Laurance, 2009[76]), they have not been well-studied in the context of solar energy.

Cumulative and population-level impacts

Little is known about how various impacts of a single solar energy facility (e.g., the combination of collision risk, habitat loss, displacement) accumulate, or the combined spatial and temporal impacts of multiple facilities on species and ecosystems. Concentrating solar energy facilities in landscapes or habitats with high irradiance could lead to significant cumulative habitat loss and collision mortality. For example, in California, US, the plurality of developments are sited in shrubland and scrublands, which are often fragile habitats with high endemism (Hernandez, Hoffacker and Field, 2014[77]). Across the US, deserts and xeric shrubland habitats are expected to be most affected by land-use change from PV and CSP deployment by 2030 (McDonald et al., 2009[78]).

Much of the literature has focused on the impacts of large utility-scale solar facilities on biodiversity, however, the cumulative impacts of multiple, smaller facilities could also pose a threat to biodiversity. In Japan and Korea, for example, medium-size solar facilities have collectively driven greater amounts of natural and semi-natural habitat loss than large-scale solar PV (approximately 66% and 86% of the overall loss in Japan and Korea, respectively) (Kim et al., 2021[79]).

In addition to the cumulative impacts across solar energy facilities, a concern is how solar impacts may accumulate with impacts from other infrastructure or human activities. A study of the combined impacts of wind and solar developments in the US, for example, indicate that these could have a significant impact on the demographics of various bird species (Conkling et al., 2022[80]) (discussed further below in 3.1.2).

3.1.2. Wind energy

Wind energy developments may consist of utility-scale turbines (>100 kilowatt [kW]) that deliver their energy to the national electricity transmission network, or small (<100kW) to medium-scale (100-500 kW) turbines that produce electricity for on-site use. Utility-scale developments involve multiple turbines connected in a wind farm, while small and medium-scale turbines for on-site use are typically installed as single units.

Wind power is produced both by onshore and offshore wind facilities. Onshore wind represents the vast majority of wind power production, with capacity growth led by China and the USA (IEA, 2020[81]). Offshore wind represents only 7% (57 GW) of total wind power capacity, with most capacity found in Europe and China (Lee and Zhao, 2022[82]). However, offshore wind is rapidly maturing and set to expand in the coming decades owing to performance and cost improvements. In 2021, 22.1 GW of new offshore installations were commissioned, representing 22.5% of all new wind installations (GWEC, 2022[83]). The expansion of bottom-fixed and floating offshore wind energy presents new risks to marine biodiversity, which remain only partially understood; however, it also presents an opportunity to reduce pressure on scare land resources and terrestrial biodiversity.

Offshore wind turbines are either bottom-fixed or floating. While bottom-fixed accounts for most offshore capacity, significant developments have been made in floating offshore in recent years. The UK, for example, installed 57 MW in 2021 (Lee and Zhao, 2022[82]). Floating offshore wind allows wind resources to be tapped in areas where water depths exceed 50-60 metres and traditional fixed-bottoms offshore installations are not viable (IEA, 2019[84]). The technology also increases the potential to harness wind energy in areas where conflicts with biodiversity and other economic activities is relatively low.

Infrastructure components of wind energy facilities include a collection of turbines (each comprising a nacelle, rotor, blades and tower), a collector sub-station, cabling that run between the substation and each turbine, and a high voltage power line connecting the substation to the power grid. In addition, fixed offshore wind facilities also have underwater components securing the turbine to the seabed (e.g., monopoles, tripods, jackets, suction caisson or gravity base) (IPCC, 2011[85]). Floating offshore wind facilities have a floating structure, typically spar buoys, spar-submersibles or tension-leg platforms (Maxwell et al., 2022[86]; SEER, 2022[87]). The structure is moored with a catenary (in the case of spar buoys and spar-submersibles) or taut-leg (for tension-leg platforms). Whilst simpler to install, catenary systems have a larger spatial footprint than taut-leg systems. Catenary mooring configurations may be anchored by drag-embedded, piled or gravity anchors. Taut-leg configurations tend to use driven or suction piles or gravity anchors. In addition to the export cable and transmission lines, other onshore infrastructure required for offshore wind facilities includes a construction port and an onshore substation (Bennun et al., 2021[1]).

Onshore wind turbines in 2019 ranged in size from 1.5-4.8 MW (IRENA, 2019[88]), however larger turbines are entering the market. Global average rotor diameter for onshore wind has increased from 82 metres (m) in 2010 to nearly 120 meters (Lee and Zhao, 2022[82]), with maximum rotor diameter of over 160 m (European Commission, 2020[89]). Average hub height has increased from 81 to 103 m during the same period (Lee and Zhao, 2022[82]). Offshore wind turbines are typically larger, with an average capacity of 8 MW and turbines as large as 15 MW now on the market. Average rotor diameter for offshore wind was 163 m in 2020 (Lee and Zhao, 2022[82]). Increases in rotor diameter and hub height enable wind farms to harness power from higher and more consistent wind speeds, increasing their efficiency. Furthermore, increased height has allowed siting of turbines in forest areas as the tree canopy has less influence on wind speed and turbulence on the higher turbines (European Commission, 2020[89]). These technological developments have potentially mixed implications for biodiversity. For example, larger more efficient wind turbines can reduce number of required turbines but increase the potential collision zone for volant species (discussed further below). Large turbines make it possible to co-locate wind turbines with forestry activities (European Commission, 2020[89]), which on the one hand could reduce land-use pressure and on the other hand make it technically possible to locate wind turbines in ecologically important forests, posing a risk to biodiversity.

Wind turbine facilities occupy a large area to allow sufficient space between turbines to reduce turbulence, follow topography and avoid obstacles. However, the area between turbines can be used for other economic or environmental purposes (see 4.3.1). When accounting for the required spacing between wind turbines, wind power is one of the most land-use intensive sources of electricity (second only to dedicated biomass), requiring on average 15 000 hectares per TWh/y. However, when accounting only for the direct footprint of wind energy infrastructure (turbine pads and access road), wind power is among the least land-use intensive of electricity sources, requiring 170 hectares per TWh/y; only nuclear and geothermal have lower land-use intensity (Lovering et al., 2022[19]).

A growing body of literature indicates that wind energy facilities can affect biodiversity in various ways, especially birds and bats for onshore wind, and birds, fish and marine mammals for offshore wind. Studies have focused primarily on the impacts of utility-scale wind developments on volant species and open habitats; less is known about impacts on other taxa. Onshore wind impacts have been better studied than offshore wind, reflecting onshore wind's longer history, wider-distribution and greater accessibility for assessment, monitoring and evaluation. For offshore wind, data and information vary considerably across geographies. For example, within Europe the knowledge base necessary for managing offshore wind and biodiversity interactions is much greater for the North and Baltic Seas than for the Mediterranean and Black Seas (European Commission, 2020[89]). The available scientific literature agrees that the key risks wind turbines pose to biodiversity are from collision mortality, habitat loss, displacement due to disturbance, barrier effects, and indirect ecosystem-level effects.

Direct mortality and morbidity

One of the most evident and measurable impacts of wind energy facilities on biodiversity is direct mortality of birds and bats, during the operation of wind energy facilities. Direct mortality or morbidity is primarily caused by collision with wind turbines. While some evidence indicates that barotrauma (tissue damage to air-containing structures caused by rapid or excessive pressure change) may also be a source of bat mortality (Baerwald et al., 2008[90]), further studies suggest it is unlikely to be significant (Rollins et al., 2012[91]; Lawson et al., 2020[92]). The data and understanding of collision risk is greater for onshore wind facilities given the technical and logistical constraints associated with assessing actual numbers of bird and bat collisions with offshore wind turbines (Hüppop, 2019[93]). Few studies show which species or species groups may be particularly vulnerable to offshore wind, and under which conditions.

Widely-cited estimates of annual bird mortality from collision in the US range from 140 000 to 679 000, based on studies from onshore wind facilities (Loss, Will and Marra, 2013[94]; Erickson et al., 2014[95]; Smallwood, 2013[96]). However, these estimates date to 2012 and wind energy capacity in the US has since increased by over 100%, and wind turbine design has changed. Mean adjusted mortality rates[4] for most US-based studies have more recently been pegged at 3-6 birds/MW/year (American Wind and Wildlife Institute, 2021[97]). With 2023 capacity in the US reaching 145 569 MW (DoE, 2023[98]), annual mortality from turbine collision in the US may be between 436 700 and 873 400 birds. (Zimmerling et al., 2013[99]) estimated average avian mortality to be 8.2 birds per wind turbine in Canada, based on observations at 43 wind farms.

Estimates in South Africa of avian mortality are slightly lower than in the US and Canada, at 2 birds/MW/year (Perold, Ralston-Paton and Ryan, 2020[100]). Collision rates with wind turbines in the Isthmus of Tehuantepec, Mexico, are higher, conservatively estimated at 9.06-12.85 birds/MW/year (although this may not be representative of national mortality rates) (Cabrera-Cruz et al., 2020[101]). Estimates of total bird mortality from wind turbine collision are orders of magnitude lower than from collisions with some other infrastructure such as building windows (Sovacool, 2009[102]; Loss et al., 2014[103]), however, mortality could have cumulative, population-level effects for certain sensitive species (see Cumulative impacts and population-level effects).

Unlike for birds, wind turbine collision is the primary source of collision mortality in bats. Bat mortality rate from collision is also higher than for birds. Hayes (2013[104]), for example, estimated that over 600 000 bats were killed in the US by wind turbines in 2012, while Smallwood (2013[105]) estimated 888 000 bat mortalities for the same period. Estimates of bat mortality tend to range from 4-7 bats/MW/year in the US (American Wind and Wildlife Institute, 2021[97]), which would correspond to ~580 000 – 1 018 983 bats per year at 2023 wind power capacity (DoE, 2023[98]). However, some estimates put average mortality much higher at around 11.6 bats/MW/year (in Canada and US) and 17.2 bats/MW/year (US only) (Smallwood, 2013[105]). In Germany, an estimated 10-12 bats are killed per turbine per year (Voigt et al., 2015[106]), equivalent to 4-4.6 bats/MW/year assuming wind turbine capacity of 2.5 MW. Mortality rates much higher than national averages have been recorded at some facilities, underscoring the importance of appropriate siting and operational practices. For example, mortality rates of 40 bats/MW/year or more have been documented at wind energy facilities in the US and Mexico (American Wind and Wildlife Institute, 2021[97]; Cabrera-Cruz et al., 2020[101]) and at old wind energy facilities in Germany, which tend to be poorly-sited and operate without curtailment (Voigt et al., 2022[107]).

Mortality from wind turbine collision has been documented in a diversity of resident and migratory bird and bat species. A global meta-study found documentation of collision mortality in 362 avian species (data covered 16 countries) and 31 bat species (data covered 12 countries) (Thaxter et al., 2017[108]). A study of 20 facilities in South Africa documented collision mortality in 130 avian species from 46 families, equivalent to 30% of the bird species identified in and around the facilities (Perold, Ralston-Paton and Ryan, 2020[100]).

While collision mortality affects a wide range of species, some species are more vulnerable to collision. Among avian species, Accipitriformes (birds of prey), Bucerotiformes (hornbills and hoopoes), Ciconiformes (storks and herons) and some Charadriiformes (shorebirds) have relatively high levels of vulnerability (Thaxter et al., 2017[108]). Small passerines tend to account for most observed mortality, but this likely reflects their greater abundance (AWWI, 2019[109]). For offshore wind facilities, modelling and empirical evidence suggests that gulls, pelicans, terns and cormorants are of greatest risk of collision (Adams et al., 2016[110]; Everaert and Stienen, 2006[111]; Skov et al., 2018[112]).

Among bat species, Nyctalus, Pipistrellus, Vespertilio and Eptesicus spp. account for the majority of mortalities in Europe (Rydell et al., 2010[113]), while in North America, the majority of bat mortalities are of hoary bat *(Lasiurus cinereus)*, eastern red bat *(Lasiurus borealis)* and silver-haired bat *(Lasionycteris noctivagans)* (American Wind and Wildlife Institute, 2021[97]). Because studies have largely focused on temperate areas (Europe and North America in particular), understanding of collision-risk for insectivorous bats is better developed than for fruit bats, which tend to live in tropical and sub-tropical zones.

Various species-specific, location-specific and facility-specific factors may contribute to a species' vulnerability. A global trait-based assessment of collision vulnerability identified migratory status, dispersal distance and habitat associations as important species-specific factors for avian vulnerability to collision (Thaxter et al., 2017[108]). The related factors of morphology (e.g. size and wing loading) (Barrios and Rodríguez, 2004[114]; de Lucas et al., 2008[115]), and flight behaviour (e.g. flight height) (Poessel et al., 2018[116]; Roemer et al., 2017[117]; McClure et al., 2021[118]) may also affect avian vulnerability (see (Marques et al., 2014[119]) for an overview). Among bat species, those adapted for open-air foraging, migratory species and tree-roosting species appear to be most vulnerable. Several studies suggest that some bat species may be attracted to turbines, although different hypotheses exist over the cause of the attraction (e.g. turbine noise, insect concentrations around turbines, or bat mating behaviour) (Richardson et al., 2021[120]; Cryan, 2014[121]; Cryan, 2008[122]). Dispersal distance has also been identified as the key species-specific factor for bat vulnerability (Thaxter et al., 2017[108]). Being nocturnal, and often high flying, bat interactions with turbines are more difficult to study; relatively little is known about why some species tend to be more vulnerable than others (Thompson et al., 2017[123]; Cryan, 2014[121]).

In addition to species-specific factors, location-specific factors and the design of wind energy facilities can affect collision risk. Location-specific factors include topography (Rydell et al., 2010[124]), overlap with migratory routes (Drewitt and Langston, 2008[125]) and other flight paths (e.g. to foraging and roosting sites) (Everaert and Stienen, 2006[126]), habitat type and quality (Heuck et al., 2019[127]) and weather conditions (Rydell et al., 2010[113]; Schmuecker et al., 2020[128]). For example, wind turbines located at avian migratory bottlenecks tend to experience particularly high rates of collision (Thaxter et al., 2017[108]), and collision risk to migratory soaring birds may increase when turbines are located along ridges that overlap with orographic lift (Marques et al., 2014[119]). Studies of bat collision mortality at North American wind facilities found that collision rates increased on nights with low wind speed, and before and after the passage of storm fronts (Arnett et al., 2008[129]).

Facility-specific factors include the configuration and scale of the wind energy facility, turbine design (e.g. size, type, and visibility), cut-in speed and hours of operation (Marques et al., 2014[119]). The impact of turbine size on collision rates is uncertain. Several studies have shown collision rates tend to be higher at larger turbines (Thaxter et al., 2017[108]; Rydell et al., 2010[124]), which could be due to the larger rotor-swept area or because the taller turbines increase the overlap with the flight heights of nocturnal, migrating songbirds and bats (American Wind and Wildlife Institute, 2021[97]; Matthew et al., 2016[130]). However, deploying fewer but larger wind turbines with greater energy output could reduce total collision risk per unit energy output (American Wind and Wildlife Institute, 2021[97]; Thaxter et al., 2017[108]). Raptor mortality on a per MW basis declined by 67%-96% (depending on the species) at the Altamont Pass Wind Resource Area after replacing smaller, low-capacity turbines with taller, higher-capacity turbines (Smallwood and Karas, 2009[131]). However, several factors could explain this decline, including the reduction in the total number of turbines, the relatively slower rotation of larger turbines, and the shift from lattice-tower turbines, which may attract raptors by providing perching sites, to modern monopole turbines (American Wind and Wildlife Institute, 2021[97]). The increasing understanding of collision risk factors is helping industry to design effective mitigation measures (Box 3.1).

While collision mortality has been most studied for birds and bats, it has also been observed for insects. One study estimates annual insect mortality from onshore wind turbines in Germany to be 1.2 trillion, equivalent to 40 million insects per turbine per year (Voigt, 2021[132]). Given documented declines in insect biomass (Hallmann et al., 2017[133]; Sánchez-Bayo and Wyckhuys, 2019[134]), further study to better understand how and to what extent wind turbines impact insect species would be justified.

For marine life, the risk of collision with offshore wind turbine foundations is likely low (Inger et al., 2009[135]). The greater risk of collision may arise from increased boat activity during surveying, construction, servicing or decommissioning of facilities. Marine mammals and turtles are at particular risk of boat collision (Maxwell et al., 2022[86]). Boat collision is a known cause of mortality and injury; it may lead to population-level effects in areas where vessel activity is already high (Rockwood, Calambokidis and Jahncke, 2017[136]). Vessel activity, and therefore collision risk, may be lower for floating offshore than for fixed offshore, as most of the construction can occur onshore (Maxwell et al., 2022[86]).

While vessel-collision risk may be lower at floating offshore wind facilities than fixed-bottom facilities, the risk of entanglement with underwater cables and mooring lines is higher (Wilson, 2006[137]; Maxwell et al., 2022[86]). Primary entanglement, where a species is entangled directly in the cables or lines themselves, has not been observed to date. The risk of primary entanglement is likely highest where catenary moorings are used because they have more slack than other mooring approaches. Secondary entanglement, where an animal gets entangled in fishing gear or other marine debris caught on cables or mooring lines, is likely to represent a greater risk than primary entanglement. Vulnerable species include those with large appendages such as humpback whales *(Megaptera novaeangliae)* and leatherback sea turtles *(Dermochelys coriacea)*, diving sea birds, elasmobranchs and fish (Maxwell et al., 2022[86]). While no evidence exists of secondary entanglement at offshore wind facilities, entanglement particularly with ghost fishing gear, is widely documented and poses a significant threat to cetaceans and other marine life (OECD, 2021[138]). Monitoring and mitigating secondary entanglement could be important as offshore wind facilities develop, as it may have a population-level impact for some species (Maxwell et al., 2022[86]).

Box 3.1. Wind energy project measures for reducing collision risks for birds and bats

Wind turbine curtailment and shut-down-on-demand

One measure to minimise collision risk of onshore and offshore wind farms during the operational phase is to slow wind turbines through feathering (angling the blades parallel to the wind) or to shut them down on demand (SDOD) when species of concern are most at risk. Curtailment can be specified for pre-defined times of day to avoid peak activity periods, ambient factors such as wind speed or temperature, which are particularly relevant for bats, or seasonality (e.g., during bird and bat migrations). Wind turbines can also be shut down on-demand when species presence is less predictable.

SDOD approaches are based on real-time observation of species activity by field observers, image-based and radar systems, or both. Shutdown can be automated and advances in machine learning and artificial intelligence (AI) may lead to further automation. SDOD has been actively used and tested on onshore wind farms but is more challenging to implement offshore. It has proven effective in reducing mortality of some avian species. For example, SDOD applied to twenty wind farms in Cadiz, Spain, reduced mortality of soaring birds by 62% and Griffon Vultures by 93%, while costing about 0.5% in energy production. In the first 5 years of radar and visual assisted turbine shutdown at the Barão de São João Wind Farm in Portugal, no mortality of soaring birds was observed during the monitoring activities (carcasses of non-targeted non-soaring birds were however detected). Furthermore, the annual average shutdown period decreased continuously from over 100 hours to only 15 hours as the shutdown procedure gained in efficiency.

Modification of wind turbine design

Another cost-effective approach that may reduce collision risk involves modifying the design of wind turbines to increase visibility to birds. Laboratory experiments have indicated that painting one of the rotor blades black may be an effective mitigation measure for collision risk. One field study in the Smøla wind farm in Norway supported these findings through a Before-After-Control-Impact which indicated that painting a single blade of each wind turbine black reduced the annual bird mortality rate by over 70% over unpainted controls, with even higher reduction of mortality for raptors. In the case of white-tail eagles, for example, no carcasses were found. However, uncertainty remains around these findings and the efficacy of painting blades as a mitigation measure. The Dutch government, together with nature organisations and private actors in the wind industry launched a follow up study in 2021, which aims to assess the impact on local birds, landscape aesthetics and aviation safety of painting blades black. The study will conclude in 2024.

Cut-in speed

The cut-in speed is the wind-speed level at which wind turbines begin turning. It has been established that bat collision and fatalities are greater at lower wind speeds. Several experiments have shown that increasing the cut-in speed of wind turbines is an effective mitigation measure to reduce bat collision fatalities. For example, one field experiment in south western Alberta, Canada, saw a 60% reduction in bat mortality when changing the cut-in speed. Similar experiments, for instance in Pennsylvania, US, have produced consistent results.

Source: (Arnett et al., 2009[139]), Effectiveness of changing wind turbine cut-in speed to reduce bat fatalities at wind facilities; (Arnett et al., 2010[140]), Altering turbine speed reduces bat mortality at wind-energy facilities, 10.1890/100103 (Baerwald et al., 2009[141]), A Large-Scale Mitigation Experiment to Reduce Bat Fatalities at Wind Energy Facilities, 10.2193/2008-233 (Bennun et al., 2021[1]), Mitigating biodiversity impacts associated with solar and wind energy development: guidelines for project developers, 10.2305/iucn.ch.2021.04.en, (Biogradlija, 2022[142]), RWE looks into black rotor wind blades for birds protection, https://www.industryandenergy.eu/renewables/rwe-looks-into-black-rotor-wind-blades-for-birds-protection; (Ferrer et al., 2022[143]), Significant decline of Griffon Vulture collision mortality in wind farms during 13-year of a selective turbine stopping protocol, https://doi.org/10.1016/j.gecco.2022.e02203 (Horn, Arnett and Kunz, 2008[144]), Behavioral Responses of Bats to Operating Wind Turbines, 10.2193/2006-465, (Huso et al., 2021[145]), Relative energy production determines effect of repowering on wildlife mortality at wind energy facilities, 10.1111/1365-2664.13853 (May et al., 2020[146]), Paint it black: Efficacy of increased wind turbine rotor blade visibility to reduce avian fatalities, 10.1002/ece3.6592 (Smallwood and Thelander, 2008[147]), Bird Mortality in the Altamont Pass Wind Resource Area, California, 10.2193/2007-032 (Smallwood and Bell, 2020[148]), Effects of Wind Turbine Curtailment on Bird and Bat Fatalities, 10.1002/jwmg.21844 (Tomé et al., 2017[149]), Radar Assisted Shutdown on Demand Ensures Zero Soaring Bird Mortality at a Wind Farm Located in a Migratory Flyway, 10.1007/978-3-319-51272-3_7.

Habitat loss and degradation

In addition to direct mortality, wind energy deployment can lead to habitat loss, species displacement and barrier effects on species. As discussed above, wind turbine facilities tend to be extensive but their direct physical footprint from the construction of permanent infrastructure (e.g., wind turbine pads, power substations and access roads) is relatively small.

Nonetheless, the construction of wind energy facilities can contribute to direct habitat loss and degradation as a result of vegetation removal (including deforestation), soil erosion and compactness, and changes in hydrology (Dhar et al., 2020[150]; Dai et al., 2015[151]; Tabassum-Abbasi et al., 2014[152]). In Scotland for example, 6 994 hectares of public forestry, amounting to 1.7% of national forests, have been cleared for wind farm development between 2000 and 2019. This figure does not account for the additional privately-owned forests cleared for wind farms (Scottish Government, 2020[153]). A study of wind turbines in Romania found a significant difference between the diversity of rare, endemic and threatened plants inhabiting areas disturbed by the turbines and surrounding undisturbed areas. The study demonstrated that plant recovery after wind energy farm construction was incomplete after ten years of low-intensity plant restoration and conservation activities (Urziceanu et al., 2021[154]).

The foundations, anchors and electrical cables of offshore wind farms alter benthic habitat, affecting primarily benthic species, but also coastal and pelagic species that depend on benthic habitat for part of their lifecycle (e.g. egg-laying) (Maxwell et al., 2022[86]; SEER, 2022[87]). Habitat loss occurs at the interface of wind energy infrastructure and the sea floor. This is a small area, but the impact could cumulate across turbines and wind energy facilities. The extent of habitat loss depends on the technology. For example, fixed-bottom turbines have a larger physical footprint on the seabed than floating wind turbines. Gravity foundations have a base up to 30 m in diameter, while jacket foundations have three cylindrical legs each of a few metres in diameter (SEER, 2022[87]).

Installation of anchors, foundations and cables can also temporarily disturb or degrade habitats. The extent and significance of these impacts depends on the technology (e.g. type of foundation, anchor and mooring line), method (e.g. whether cables are buried or protected), and local conditions (e.g. sediment type) (Maxwell et al., 2022[86]). In some cases, disturbance may continue throughout the project lifetime. For example, catenary mooring lines may drag across the sea floor. Unlike for habitat loss, the seabed can naturally recover (at least partially) from disturbance (Maxwell et al., 2022[86]). The rate of recovery is typically faster for shallower water with soft bottoms as they are more dynamic (SEER, 2022[87]).

Habitat fragmentation and barrier effects

The construction of wind energy facilities can fragment habitats, leading to a host of impacts (see discussion under solar energy). While the spatial footprint of wind turbine infrastructure such as access roads may not be large, they can have a disproportionate impact on habitat connectivity. A study of 39 wind facilities, for example, found that new facilities decreased the amount of undeveloped land by 1.8% but changed metrics of landscape pattern from 50-140% (Diffendorfer et al., 2019[155]). The extent of pre-construction development can play a key role in determining overall impact of a wind energy facility. Utilising existing development reduced habitat fragmentation.

Wind turbines can also provide a barrier effect to species if they are located on paths used to access foraging or breeding grounds, or for migration. Various species of birds have been observed to change their flight paths to avoid flying through wind energy facilities (Cabrera-Cruz and Villegas-Patraca, 2016[156]). The impact of barrier effects is likely greatest on migratory species, as adjusting flight paths to avoid (often multiple) wind energy facilities (and potential rest stops), may burn scarce energy reserves. Barrier effects for non-volant species may occur if the wind energy facility is fenced.

Habitat alteration or creation

The new substrate provided by turbine foundations can also create new habitat for fish and marine invertebrates (ter Hofstede et al., 2022[157]). Several studies have documented the establishment of communities on offshore turbine foundations (Lindeboom et al., 2011[158]) (De Mesel et al., 2015[159]). Turbines have also been proved to provide refuge for fish (Bergström, Sundqvist and Bergström, 2013[160]; Langhamer, 2012[161]). In the Netherlands' North Sea, the Flat Earth Consortium pilot launched by the Ministry of Economic Affairs of the Netherlands and the World Wide Fund for Nature (WWF) aims at further enhancing this effect by adding artificial reef structures in areas around turbine foundations that have been damaged by trawling (Didderen, 2019[162]). The reef effect can, however, be both positive and negative for biodiversity. It is important to consider the historical context (i.e., has there been past loss of hard substrate habitats), carefully assess impacts on communities and ecosystem services, and monitor impacts (e.g., potential colonisation by invasive non-native species). Furthermore, it is necessary to consider the long-term implications of decommissioning infrastructure (e.g., whether infrastructure supporting new reef structures should be removed) (European Commission, 2020[89]).

Behavioural changes, physiological changes and species displacement

Species respond differently to the physical presence and noise of wind energy facilities. Some are unaffected, some appear to be attracted to the facilities, while other species avoid wind energy facilities (Łopucki, Klich and Gielarek, 2017[163]; Lindeboom et al., 2011[158]). Non-species-specific factors, such as season (Peschko et al., 2020[164]), wind-turbine location and extent of land-use change (Fernández-Bellon et al., 2018[165]), may affect the magnitude and direction of species responses.

Species displacement (effectively habitat loss), owing to avoidance of wind energy facilities may have a more extensive impact on biodiversity than direct habitat loss through clearance. According to one estimate, direct habitat loss through clearing only accounts for 3-5% of wind turbines' impact area, while most of their impact area is due to habitat fragmentation and species avoidance behaviour (McDonald et al., 2009[78]).

Avoidance of onshore wind energy facilities has been observed for a number of species, and is particularly well-documented in birds and bats (Fernández-Bellon et al., 2018[165]) (Shaffer and Buhl, 2015[166]) (Pearce-Higgins et al., 2009[167]) (Pearse et al., 2021[168]; Therkildsen et al., 2021[169]). A study of 130 black kites *(Milvus migrans)* at the migratory bottleneck of the Strait of Gibraltar found that the birds avoided areas up to approximately 674 m from operating wind turbines and an estimated 3-14% of suitable soaring area (Marques et al., 2019[170]). Another study found that activity within 1 000 m of wind turbines by gleaners and fast-flying bats is reduced by 54% and 20%, respectively (Barré et al., 2018[171]). A multi-site study of several avian species in the UK suggested that displacement impacts during construction may be as or even more important than displacement triggered by turbine operations (Pearce-Higgins et al., 2012[172]).

Although less-studied, evidence of species avoiding operating wind energy facilities also exists for non-volant terrestrial mammals, such as pronghorn (*Antilocapra americana*) in the US (Smith et al., 2020[173]; Milligan et al., 2023[174]), reindeer (*Rangifer tarandus*) in Sweden (Skarin, Sandström and Alam, 2018[175]), European hare (*Lepus europaeus*) and roe deer (*Capreolus capreolus*) in Poland (Łopucki, Klich and Gielarek, 2017[163]), wolves (*Canis lupus signatus*) in Portugal (Skarin, Sandström and Alam, 2018[175]; Łopucki, Klich and Gielarek, 2017[163]), blackbuck (*Antilope cervicapra*), chinkara (*Gazella bennettii*), golden jackal (*Canis aureus*) and jungle cat (*Felis chaus*) in India (Kumara et al., 2022[176]). Subsoil biodiversity can also be affected by wind energy facilities. For example, in the Netherlands, earthworms were less abundant close to wind turbines, likely due to their avoidance of vibratory noise associated with wind turbine operation (Velilla et al., 2021[177]).

Offshore wind facilities have also been observed to affect species behaviour. The construction phase is when most displacement of marine mammals can be observed, owing to pile-driving noise (Box 3.1). Displacement of harbour porpoises *(Phocoena phocoena)* (Dähne et al., 2013[178]; Brandt et al., 2018[179]; Graham et al., 2019[180]) and harbour seals *(Phoca vitulina)* (Russell et al., 2016[181]), have been well-documented. However, other cetaceans and pinnipeds as well as fish are also likely affected by construction noise (Kok et al., 2021[182]; Bailey, Brookes and Thompson, 2014[183]).

The operation phase of offshore wind turbines can displace bird species, similar to onshore wind turbines (Lindeboom et al., 2011[158]; Peschko et al., 2020[164]). Displacement of non-avian species owing to offshore wind turbine operations has been less studied. A seven-year monthly demersal trawl survey using a Before-After-Control-Impact (BACI) design found that relative decreases in fish and invertebrate abundances during wind farm operation were neither statistically nor substantively evident. Some species (e.g., black sea bass) had significantly higher catch per unit effort, consistent with the artificial reef effect (Wilber et al., 2022[184]).

Species displacement can be temporary (e.g., during construction or while species become accustomed to the operation of wind turbines), or it could persist throughout the lifetime of the wind energy facility. The impact of temporary displacement is likely to depend on when it occurs (e.g. impacts could be particularly high during breeding season) (Pearce-Higgins et al., 2012[172]). Accounting for seasonal and daily movement patterns of species of concern in the construction schedule (e.g., scheduling road construction or underwater piling outside breeding seasons) could help reduce construction phase impacts.

In addition to avoidance behaviour, species may experience higher levels of stress around wind energy facilities. For example, roe deer close to wind energy facilities were found to have higher levels of the stress hormone cortisol (Klich et al., 2020[185]), while wolves close to wind energy facilities were found to have lower reproduction rates (Ferrão da Costa et al., 2017[186]). Male Japanese tree frogs living close to wind turbines were found to have a higher call rate and corticosterone levels and lower immunity than frogs from control sites without turbines during the breeding season (Park and Do, 2022[187]).

Knowledge gaps remain regarding the behavioural and physiological responses of species to wind turbine facility construction and operation, and how this effects ecosystem structure and function. In addition to continued study of volant species responses, further studies of the responses of non-volant terrestrial and marine species, in particular fish and invertebrates,[5] and a variety of ecosystems[6] could help create a fuller picture of wind energy impacts and help inform mitigation.

Box 3.2. Minimising underwater noise impacts

One of the main impacts of offshore wind farms is species disturbance from underwater noise, especially in the construction phase.

Piling protocols

Piling activities are associated with the highest level of underwater noise in the construction phase of offshore wind farms. In the United Kingdom, good practices to reduce risks of injury or death to marine mammals due to piling operations have been developed as the 'piling protocol' by the Joint Nature Conservation Committee (JNCC). It involves implementing a mitigation zone of 500 m radius from the piling location, in which presence of marine mammals is monitored. Piling activities should be delayed in case of detection. Piling power should also be gradually ramped up during a minimum of 20 minutes so that marine animals can move away from the noise source and avoid harmful exposure.

The piling protocol has been used and adapted to many wind farm projects. The TEESSIDE offshore wind power project off the coast of Redcar in England was developed in an area where cetacean species protected under UK law were present (e.g., harbour porpoise, bottlenose dolphin, minke whale) and a strict protocol to ensure that the area of piling activities was clear of marine mammals was implemented. Prior to each piling exercise, a vessel circled 250 m around the piling site. A hydrophone was also used to listen for marine mammal vocalisations. Piling work was authorised to start once the area had been monitored for 30 minutes and no marine mammals had been detected.

Acoustic deterrent devices (ADDs)

ADDs are used to warn species away from dangers such as fishing gear. They could also be used to create a temporary safety exclusion zone around turbines during their constructions. The JNCC piling protocol recommends the use of ADDs to avoid underwater noise impacts during the construction of offshore wind farms, but its effectiveness has only been tested on a few species. For example, a study found effective deterrence for minke whales. The JNCC has published a review of existing ADD technologies and the evidence-based of their effectiveness on different species.

Source: (McGarry et al., 2022[188]), Evidence base for application of Acoustic Deterrent Devices (ADDs); (McGarry, 2017[189]), Understanding the Effectiveness of Acoustic Deterrent Devices on Minke Whale (Balaenoptera acutorostrata), A Low Frequency Cetacean; (Bennun et al., 2021[11]), Mitigating biodiversity impacts associated with wind and solar energy development; (JNCC, 2010[190]), Statutory nature conservation agency protocol for minimising the risk of injury to marine mammals from piling noise.

Impacts from invasive alien species

Ecological communities may be affected at wind energy facilities by the introduction or spread of invasive alien species (IAS). The movement of people, construction material and equipment provide a vector for IAS. For example, in the case of offshore wind facilities, increased vessel activity for construction and maintenance could introduce non-native species to wind facility sites. Marine vessels are well-known vectors of IAS, which they transport via hull fouling and ballast water exchange (Costello et al., 2022[191]). Additionally, the degradation or alteration of habitats from wind turbine construction could make ecosystems more susceptible to the spread of non-native and invasive species. For example, operational monitoring at the Serra da Lousã onshore wind facility in Portugal found two new IAS present. Two other IAS, already present at the site, had spread along access roads and turbine pads (Passo, 2017[192]). Offshore wind turbines require anchors, pilings or foundations that can alter benthic habitats, providing a potential niche and stepping-stone for invasive alien species (Adams et al., 2014[193]).

Ecosystem services impacts

Wind energy facilities potentially affect the quantity or quality of provisioning, regulating, supporting and cultural ecosystem services, or people's access to these services. A commonly reported concern is the impact of wind turbines on cultural values. The development of wind energy facilities can undermine the aesthetic value of landscapes, affecting land value and impacting tourism (Gibbons, 2015[194]; Dröes and Koster, 2021[195]; Voltaire and Koutchade, 2020[196]; Parsons et al., 2020[197]). For example, the installation of offshore wind facilities in Catalonia, Spain could affect preferences for beach trips by residents and tourists and result in a loss of welfare (Voltaire and Koutchade, 2020[196]). Although declines in tourism could reduce pressure on other ecosystem services, underscoring the potential for trade-offs across ecosystem services.

The land-use change associated with wind farm deployment can also undermine an ecosystem's ability to sequester and store carbon. For example, the construction of 3 848 wind turbines in Scotland led to 4.9 million tonnes of CO_2 emissions because of land use change. Emissions vary from turbines across Scotland vary from 16 g CO_2/kWh in shrubland to 1 760 g CO_2/kWh in peatland (Albanito et al., 2022[198]).

Where wind energy is deployed in agricultural land or fishing grounds, it may negatively affect provisioning services. For example, in some jurisdictions, fisheries cannot or can only partly access offshore wind farms. Some offshore wind facilities overlap with existing fishing grounds, presenting an opportunity cost. For example, 90% of Danish and 40% of German annual plaice landings in the German EEZ overlap with areas where offshore wind farms are and will be developed (Van Hoey et al., 2021[199]). International gillnet fisheries could lose up to 50% in landings within the North Sea German EEZ. In theory, part of the loss to fisheries could be compensated by fishery displacement (see indirect impacts) and, in the longer term, potential spill over effects, where offshore wind facilities act as marine protected areas allowing populations to thrive and then spill over to surrounding areas where they could bolster fishing (Van Hoey et al., 2021[199]; Ashley, Mangi and Rodwell, 2014[200]).

Cascading impacts

Reduced prevalence of some species in and around wind energy facilities may have extensive, albeit poorly understood knock-on effects on ecological communities (Keehn and Feldman, 2018[201]). Long-term monitoring is required to identify and understand the extent of these impacts. One mechanism by which knock-on effects can occur is cascading impacts, whereby declines in species in higher trophic levels affects those in lower trophic levels. For example, a long-term study in the Western Ghats, India, found that both the abundance of predatory birds and the frequency of predation attempts by raptors on ground-dwelling prey were almost four times lower in sites with wind turbines (Thaker, Zambre and Bhosale, 2018[202]). The density of the endemic superb fan-throated lizard (Sarada superba) was significantly higher in sites with wind turbines. Furthermore, the lizards in wind sites showed physiological, behavioural and morphological differences consistent with the effects of predator release. A study of bat mortalities in wind turbines in Germany concluded that may lead to the loss of trophic interactions and ecosystem services, contributing to functional simplification and impaired crop production, respectively (Scholz and Voigt, 2022[203]).

Indirect impacts

The construction and operation of wind energy facilities may have indirect impacts by displacing other activities. For example, a 25-fold increase in offshore wind energy is expected in the North and Baltic Seas by 2050, potentially competing with fisheries and aquaculture. Fishermen tend to respond by displacing fisheries or by switching fishing gear (e.g., mobile gear to crab or lobster potting) (EC, 2021[204]). This could increase fishing pressure elsewhere, with potential implications for biodiversity.

Cumulative impacts and population-level effects

The cumulative impacts and population-level effects of multiple wind energy developments (and other human pressures) remain understudied and poorly understood. While current levels of mortality from wind turbine collision are thought to have negligible impacts on the populations of many of the affected species, some species could face significant cumulative impacts and population-level effects. Migratory birds and bats are likely to be at particular risk of cumulative impacts as they may cross the paths of multiple wind farms. Furthermore, bats, and some of the groups of bird species demonstrating the highest risk of collision (e.g. Accipitriformes, Bucerotiformes, Circoniformes), tend to be species with long lifespans, low fecundity and late ages of maturity, which makes them particularly sensitive to additional mortality (Thaxter et al., 2017[108]).

A recent assessment of the vulnerability of populations of 23 priority bird species killed at wind and solar facilities in California, USA, concluded that 48% were vulnerable to population-level effects from additional mortalities caused by renewables and other sources. Notably, the study found that Californian renewable energy facilities could have population-level effects not only for local subpopulations but also for non-local subpopulations (e.g. distant subpopulations of migratory species) (Conkling et al., 2022[80]).

Based on expert elicitation and population models, (Frick et al., 2017[205]) concluded that wind energy development could have significant population-level impacts on migratory bats in North America, with the hoary bat population declining by as much as 90% in the coming 50 years. The impact on populations of rare and already threatened species are also of particular concern. For example, the little brown myotis *(Myotis lucifugus)*, which was listed as Endangered in 2014 under the Species At Risk Act (SARA), accounted for 13% of all mortalities from wind turbines in Canada and 87% of mortalities in Ottawa (Zimmerling and Francis, 2016[206]).

A global review of bat multiple mortality events (MMEs), defined as cases in which ≥ 10 dead bats were counted or estimated at a specific location within a maximum timescale of a year, identified collision with wind turbines as the leading cause of reported MMEs in bats, alongside white-nose syndrome (O'Shea et al., 2016[207]). Owing to their important role in ecosystem function and ecosystem service provision (see Box 3.3), the cumulative impact on bats of turbine collision, white-nose syndrome (for North American bats), and other threats such as habitat loss, hunting for bushmeat and climate change is a conservation concern (Frick, Kingston and Flanders, 2019[208]).

A growing body of literature examines population-level impacts and identifies species whose populations are most likely to be significantly affected (see e.g. (Diffendorfer et al., 2021[209])). Long-term studies of wind turbine collisions and of population dynamics (particularly for bats, as these are less well understood) are critical for improving understanding of population-level impacts. Cumulative impacts of offshore wind energy facilities also warrant further exploration (see (Brignon et al., 2022[210]) for priority research areas).

Box 3.3. The ecological and economic importance of bats

Bats are the second most diverse and abundant order of mammals and play essential roles in many ecosystems across all continents except Antarctica. They provide a wide range of ecosystem services such as seed dispersal, pollination, regulation of agricultural pests, and material and nutrient distribution. For example, bats are responsible for pollinating hundreds of plant species, including economically important plants such as bananas, mangos, agave and durians, some of which rely primarily or exclusively on bats for pollination. The total global value of bats in pollination services was estimated at USD 200 billion in 2005, which represented 9.5% of the world food crop production value. Bats are also crucial for maintaining and regenerating tropical forests through seed dispersal. As important predators of various insect species, bats reduce the need for pesticide use. In the US alone, this service is estimated to save USD 3.7-53 billion per year in pesticide costs.

Despite their importance, over a third of bat species assessed by the International Union for Conservation of Nature are considered threatened or data deficient. Over half of the species have unknown or decreasing population trends.

Source: (Kasso and Balakrishnan, 2013[211]), Ecological and Economic Importance of Bats (Order Chiroptera), 10.1155/2013/187415; (Frick, Kingston and Flanders, 2019[208]), A review of the major threats and challenges to global bat conservation, 10.1111/nyas.14045.

3.1.3. Electricity networks (transmission and distribution lines)

The electricity system relies on transmission and distribution lines to transfer power from the source to the end-user. Transmission systems are designed to transfer higher voltages over long distances, from the site of energy generation to a substation. Distribution lines carry lower voltages over shorter distances from the substation to end-users (Biasotto and Kindel, 2018[212]). Existing power line networks consist predominantly of overhead lines, but also of underground cables.

A double trend is increasing the need for long-distance power transfers and the associated power line infrastructure. First, deeper market integration with increasing interregional and international trade in electricity implies that greater amounts of electricity have to be transported across large distances (van der Weijde and Hobbs, 2012[213]) (Pollitt, 2009[214]). This is especially the case in the European market where market liberalisation encourages cross-border electricity transfers (Brown, 2015[215]). Second, the increased use of renewable sources of electricity worldwide requires soothing the allocation of power across large areas, as such sources are often located in remote areas far from load centres and as the availability of wind and solar in particular is intermittent (van der Weijde and Hobbs, 2012[213]). Weather-dependent electricity also increases situations where excess local generation must be exported to balance demand and supply across regions (Brown, 2015[215]).

The IEA Net Zero pathway estimates that investment in electricity networks will need to increase to USD 820 billion/year by 2030 from about USD 260 billion/year in 2021 (IEA, 2021[216]). This would ensure the provision of large amounts of power line infrastructure as well as the modernisation of existing assets. Europe for instance, already houses 300 000 km of transmission lines and 10 million km of distribution lines (similar scales are found in the US) (IEA ETSAP, 2014[217]), and the EU has set the target of increasing the interconnectivity of national power grids by 15% by 2030 requiring an estimated 44 700 km of new or refurbished transmission lines (EU, 2019[218]).

While networks consist predominantly of overhead lines, underground and subsea cable technology is increasingly being deployed. From 2010-14, 8 000 km of subsea and underground lines were installed globally (Europacable, 2021[219]).

The scientific literature and the body of environmental impact studies evaluating the potential impacts on biodiversity of transmission and distribution lines are well developed, particularly for birds. However, knowledge gaps remain, particularly for impacts on non-avian species (e.g., amphibians), habitats and ecosystems (Biasotto and Kindel, 2018[212]). The scientific literature mainly covers the impact of overhead transmission lines, with fewer studies examining the impacts of distribution lines, which constitute a significantly larger network, and underground cables (Bernardino et al., 2018[220]).

Direct mortality and morbidity

One of the direct impacts of power lines is bird mortality caused by collision and electrocution. Collision risk is primarily associated with transmission lines, while electrocution risk is greatest for distribution lines. Overall, power lines are estimated to kill hundreds of thousands to millions of birds every year, including threatened species (Loss, Will and Marra, 2014[221]; Loss, Will and Marra, 2015[222]; Rioux, Savard and Gerick, 2013[223]). Several studies indicate that mortality induced by power line collision can have significant population-level impacts for some species (Schaub and Pradel, 2004[224]; Schaub et al., 2010[225]; Loss, Will and Marra, 2012[226]).

Some species are particularly prone to collision with power line. Species-specific factors such as type of vision, migratory behaviour and flight behaviour influence collision risk (Bernardino et al., 2018[220]). For example, species with high wing loading (i.e., weight to wing area ratio), such as bustards, storks, eagles, vultures and cranes, are at higher risk due to their low manoeuvrability. Site-specific factors also influence collision risk. Power lines that are placed perpendicular to major migratory corridors pose high risks for species on migration (Shobrak, 2012[227]). Similarly, power lines that cross important bird habitats such as wetlands and coastal areas are assumed to pose higher collision risks (Adrushchenko and Popenko, 2012[228]). Finally, pylon and wire characteristics are thought to influence power line bird collisions. There is a general agreement that wire height, number of levels of wire, wire diameter and presence of earth wires may influence bird collision rates (Bernardino et al., 2018[220]). However, very few studies manage to isolate one specific characteristic and its correlation with bird mortality.

Electrocution occurs predominantly at low to medium-voltage power lines. High voltage power lines tend not to have live and earthed components sufficiently close for an animal to touch both at once. The technical design of electricity transmission and distribution infrastructure (e.g., distance between exposed wires and other energised or grounded elements, and how insulators are attached to cross-arms), is a key determinant of risk (RPS, 2021[229]). Birds that use pylons for nesting or foraging are vulnerable, particularly if they are medium-sized or large birds likely to touch multiple exposed elements (Guil et al., 2011[230]; Tintó, Real and Mañosa, 2010[231]; Lehman, Kennedy and Savidge, 2007[232]).

Electrocution has been identified as the main cause of population decline for several avian species (Biasotto and Kindel, 2018[212]). It is of particular concern for threatened species and long-lived species. In Iran, 15% of the 235 birds reported to be electrocuted in 2018 were species of conservation concern, including the steppe eagle *(Aquila nipalensis)* and the Egyptian vulture *(Neophron percnopterus)* (Kolnegari et al., 2020[233]). In Southern Europe, electrocution poses risks to the viability of populations of Bonelli's eagle *(Aquila fasciata)*, even though mortality rates from power lines are low (Hernández-Matías et al., 2015[234]).

While the literature on electrocution from power lines focuses particularly on avian electrocutions, power lines also electrocute other taxa, including mammals and reptiles (e.g., snakes). Electrocutions of various primate species across Asia, Africa and Latin America have been documented (Katsis et al., 2018[235]). Among these species are threatened species, such as the critically endangered Javan slow loris *(Nycticebus javanicus)* (Moore, Wihermanto and Nekaris, 2014[236]) and the western purple-faced langur *(Trachypithecus vetulus nestor)* (Moore, Nekaris and Eschmann, 2010[237]; Parker, Nijman and Nekaris, 2008[238]). Electrocution has been a principal mortality factor for the endangered Central American squirrel monkey subspecies *Saimiri oerstedii citrinellus* and *Saimiri oerstedii oerstedii* (Boinski et al., 1998[239]).

Knowledge gaps remain concerning risk factors for collision and electrocution. Furthermore, the evidence base is geographically skewed. According to a recent review (Guil and Pérez-García, 2022[240]), most (80%) studies of bird electrocution at power lines are from developed countries, mostly in Europe and North America. No systematic studies exist for Oceania, and few exist for South America and Africa. Additionally, while evidence for the effectiveness of some mitigation measures exists, many widely accepted mitigation measures have not been consistently tested or have yielded very different results (Biasotto and Kindel, 2018[212]). Examples of mitigation measures and their evidence base is provided in Box 3.4.

Box 3.4. Potential mitigation measures to reduce collision and electrocution risk

Routing/re-routing of power lines

Collision risk can be partly mitigated through optimal power line routing. For example, siting power lines parallel to migratory routes, rather than crossing them, and avoiding areas where birds congregate, such as wetlands, can help avoid collisions.

Burying power lines

Undergrounding of low- and medium voltage networks is effective in reducing collision and electrocution risk. It may be an appropriate solution in certain areas (e.g., migratory bird corridors). However, undergrounding can face technical, economic and legal challenges and has therefore been used sparingly, particularly in the case of high voltage lines. While burying power lines can reduce electrocution and collision risk, it is important to consider other potential ecological impacts from burying power lines, as these impacts may also be significant in sensitive habitats.

Fitting of rejecters and insulation to reduce electrocution risk

Suitably arranged rejecters can deter birds from perching on poles and towers, thereby reducing the risk of electrocution. These can be retrofitted to dangerous arrangements or used on new equipment to ensure that birds do not perch in areas where they might be at risk. Insulation sheathing, hoods or plastic caps may be retrofitted or factored into the design of new infrastructure, to increase the distance between exposed conducting cables/wires and earth sources. Retrofitting of insulation on pylons in Mongolia reduced raptor mortalities by an estimated 85%.

Reconfiguration of transmission lines

Reconfiguration could involve increasing thickness and visibility of wires and decreasing their span. For example, a study of frugivorous bats in Sri Lanka suggests that orienting wires horizontally rather than vertically can reduce electrocution risk for large bat species. However, further research is required to assess the effectiveness of reconfiguration for reducing mortality.

Markers

The use of markers (bird diverters) to increase the visibility of power lines and earth wires can reduce collision mortality for some bird species by around 50%. Markers may not be effective or sufficient for all species. In Austria and Hungary, a combination of burying some power lines and marking others with diverters significantly reduced mortality of great bustards *(Otis tarda)*. In Lithuania, monitoring undertaken as part of the EU LIFE-funded project "Installation of the bird protection measures on the high voltage electricity transmission grid in Lithuania" found annual mortality of 3.6 birds/km of high voltage electricity lines where visibility increasing measures had been installed, compared to 11.1 birds/km of lines without visibility measures.

Habitat management

Considerable opportunities exist for ecological enhancement through the management of habitats beneath and around power lines to benefit wildlife. There may be opportunities to mitigate (or compensate) for losses elsewhere, through managing habitats to benefit species that are not considered at risk of collision. Habitat management to deter large raptors might be partially successful if it reduces attractiveness of the areas to these birds. On the other hand, power poles and towers offer perching, roosting and nesting sites for some large birds. Bird-safe power lines enable raptors, storks, ravens and other birds to nest in otherwise treeless landscapes, which may benefit these species.

Source: (Bernardino et al., 2018[220]), Bird collisions with power lines: State of the art and priority areas for research, 10.1016/j.biocon.2018.02.029; (Chevallier et al., 2015[241]), Retrofitting of power lines effectively reduces mortality by electrocution in large birds, 10.1111/1365-2664.12476; (Dixon et al., 2018[242]), Efficacy of a mitigation method to reduce raptor electrocution at an electricity distribution, https://conservationevidencejournal.com/reference/pdf/686; (Haas et al., 2005[243]), Protecting birds on powerlines: a practical guide on the risks to birds from electricity transmission facilities and how to minimise any such adverse effects; (Jenkins, Smallie and Diamond, 2010[244]), Avian collisions with power lines: a global review of causes and mitigation with a South African perspective, 10.1017/s0959270910000122; (LOD, 2019[245]) (Raab et al., 2011[246]), Underground cabling and marking of power lines: Conservation measures rapidly reduced mortality of West-Pannonian Great Bustards, 10.1017/s0959270911000463; (Raptor Protection of Slovakia, 2019[247]), Protecting birds from power lines focusing on countries of Danube/Carpathian region; (RPS, 2021[229]), Electrocutions and Collisions of Birds in EU Countries: The Negative Impact and Best Practices for Mitigation.

Habitat loss and degradation

Installing and maintaining overhead power lines can require clearing and managing vegetation in the zone situated below the cables (the right of way - RoW), to avoid interference and risks to the cables. The extent of habitat loss and degradation depends on the location and sensitivity of the affected habitat, the size of the project (e.g., higher voltage cables tend to require larger RoW) and the design of the project. Although generally of a small width, the RoW constitutes linear habitat loss or degradation over hundreds of kilometres. Furthermore, destroying or degrading habitat under power lines could have wider ecosystem effects by affecting hydrological regimes and exacerbating erosion. Studies of habitat loss and degradation from power line construction are, however, scarce. Habitat fragmentation and species avoidance (effective habitat loss) have been better studied for power lines (Biasotto and Kindel, 2018[212]).

Habitat fragmentation and barrier effects

RoWs can fragment habitats, leading to edge effects. Clearance of forest for power lines, for example, has been found to create microclimatic edge gradients qualitatively like those observed at forest edges adjacent to larger clearing (Pohlman, Turton and Goosem, 2009[248]). These microclimatic edge gradients can negatively impact interior forest plant species, while favouring disturbance-adapted plants, thereby leading to changes in ecological communities (Pohlman, Turton and Goosem, 2009[248]; Prieto et al., 2013[249]). Changes in fruiting phenology of species have also been observed at linear edges in the Atlantic Forest, with potential consequences for plant-seed disperser interactions (Reznik, Pires and Freitas, 2012[250]).

For small arboreal mammals, such as the lemuroid ringtail possum (*Hemibelideus lemuroides*), the loss of canopy connectivity resulting from power line construction in forests significantly limits their movements and could therefore fragment their population (Goosem and Marsh, 1997[251]; Wilson, Marsh and Winter, 2007[252]). Other taxa, such as salamander, may also be negatively impacted through the barrier effect created by power lines (Cecala, Lowe and Maerz, 2014[253]).

Habitat alteration or creation

While habitat loss and fragmentation are generally negative for biodiversity, in some contexts power line deployment can benefit native and threatened species by creating new habitats. For example, various bird species, including threatened species, use pylons for nesting and foraging (Biasotto and Kindel, 2018[212]; Arkumarev et al., 2014[254]). Others, such as scrub-shrub birds have been found to successfully nest in RoW (King et al., 2009[255]). Use of RoW may allow some species to increase their range and population size (Dixon et al., 2013[256]; Howe, Coates and Delehanty, 2014[257]), although it may increase electrocution risk.

Non-avian taxa may also benefit from RoWs. For example, rare grassland species in North America and the gastropods (e.g. snails and slugs) that depend on them have been found to benefit from the clearance and maintenance of vegetation in RoWs (Nekola, 2012[258]). The open habitats created by power line construction and operation in Central Europe supports relatively high insect diversity (Plewa et al., 2020[259]). They also provide a corridor for certain species, such as large carnivores, to move across the landscape (Clarke, Pearce and White, 2006[260]; Bartzke et al., 2014[261]; Smith et al., 2008[262]). Positive impacts may occur when RoWs facilitate movement of native species across the landscape. Large carnivores, for example, have been found to prefer moving along RoWs (Bartzke et al., 2014[261]; Smith et al., 2008[262]). These mechanisms that benefit native and threatened species may also benefit invasive alien species and lead to their spread (see discussion below).

Whether the net benefit for biodiversity is positive or negative depends on the local context. (Eldegard, Totland and Moe, 2015[263]) suggest that management plans for power lines should differentiate between clearing through high conservation value forests where edge effects should be limited and clearings that can act as replacement habitat.

Behavioural changes, physiological changes and species displacement

The presence of power lines may lead to species displacement (effective habitat loss) or provide a barrier to species movement. For example, displaying male bustards were found to reject sites within 350-400 m to medium voltage power lines. Relative rejection of potential displaying sites was estimated to be up to 3 500 m from power lines (Lóránt and Vadász, 2014[264]). Other studies have shown grassland bird species avoiding areas up to 6 000 m from power lines (Pruett, Patten and Wolfe, 2009[265]; Gillan et al., 2013[266]), and concluded that the presence of power lines contribute to abandonment of certain territories by Bearded Vulture (Krüger, Simmons and Amar, 2015[267]). Nesting and roosting sites (Santiago-Quesada et al., 2014[268]; Silva et al., 2010[269]) and demographic rates (e.g. nest survival, recruitment and population growth) (Gibson et al., 2018[270]) of some bird species have also been found to be negatively influenced by power lines. Avoidance behaviour has also been documented in ungulates, such as reindeer and white-tailed deer (Bartzke et al., 2014[261]; Rieucau, Vickery and Doucet, 2009[271]; Nellemann et al., 2001[272]).

In addition to their physical presence, power lines can affect species through noise and electromagnetic fields (EMF). The potential disturbance through noise during the construction phase of power lines are mentioned in several environmental impact statements for power line developments (Biasotto and Kindel, 2018[212]). (Colman et al., 2015[273]) suggest that disturbance through noise may temporarily reduce the presence of reindeer around the construction site. However, during the operational phase, noises associated with clicks or energy discharges in power lines are limited and their effect on wildlife is uncertain.

While impacts of EMF have been observed, relatively little is known about their extent and significance (Biasotto and Kindel, 2018[212]). Observed impacts of EMF include physiological impacts, such as altered development, behavioural effects (e.g. attraction, avoidance), and impaired navigation and orientation (Farr et al., 2021[274]; Biasotto and Kindel, 2018[212]). For terrestrial species, evidence exists of reproductive effects in birds (Fernie and Reynolds, 2005[275]; Tomás et al., 2012[276]), behavioural changes in ruminants

(Burda et al., 2009[277]) and physiological changes in plants (e.g. increased genetic mutations and altered enzymatic activity) (Aksoy, Unal and Ozcan, 2010[278]; Mahmood et al., 2013[279]). In the marine environment, elasmobranchs, crustacea, cetacea, bony fish, and marine turtles, have been shown to be sensitive to EMF (Copping and Hemery, 2020[280]; Gill et al., 2014[281]). For example, a recent study of two commercially important lobster species *(H. gammarus and C. pagurus)* concluded that EMF emissions from subsea power cables could have a measurably affect their early life history and consequently population dynamics (Harsanyi et al., 2022[282]).

Owing to the rapid growth in offshore wind, further research into the EMF impacts from power lines in the ocean would be beneficial. Floating offshore wind may require long distance cables transmitting high voltage direct current, which typically emits high intensity magnetic fields over a greater spatial scale compared to alternating current.

Impacts from invasive alien species

The degradation of habitats or alteration of habitats resulting from power line construction and maintenance of RoW can create conditions for invasive alien species to colonise and disperse along RoW corridors (Biasotto and Kindel, 2018[212]). As with solar and wind energy, few studies examine the role of power lines in facilitating invasive alien species. The extent to which power lines support IAS is likely to be context specific. A study in Quebec, Canada concluded that power lines were efficient dispersal vectors of invasive plant species in fen but not bogs (Dubé, Pellerin and Poulin, 2011[283]). A Finnish study found that power line sites that supported alien species were characterised by productive soils light, surrounded by a dense urban fabric (Lampinen, Ruokolainen and Huhta, 2015[284]).

Pylons that provide perching, roosting or nesting sites for birds may facilitate spread of invasive plans species as these are also defecation sites. A study of black cherry *(Prunus serotines)*, an invasive plant species in agricultural land in Europe, found high abundance under electricity pylons. A survey of 124 areas under pylons and 124 paired control plots in agricultural land found the plant in 82% of electricity pylons and 2% in control plots. The land under the pylons was relatively untouched compared to intensively managed agricultural land and therefore provided a refuge for the plant (Kurek, Sparks and Tryjanowski, 2015[285]).

Indirect impacts

The construction of power lines could facilitate access to previously unexploited areas, potentially threatening biodiversity. Indirect impacts from the construction of linear infrastructure, particularly roads, are well-documented albeit not often quantified (Barber et al., 2014[286]; Richardson et al., 2017[287]). While evidence of the induced impacts of power lines is slim, studies hypothesise that their construction can lead to increased deforestation (Hyde, Bohlman and Valle, 2018[288]) and hunting (Bartzke et al., 2014[261]).

Another indirect impact that can occur from power line construction and operation is increased fire risk and associated habitat loss. This is of particular concern given the increasing risk of forest fires under climate change (OECD, 2023[289]). Power line construction in forests could alter the vegetation community structure and composition at the edges of the utility corridor, and alter the microclimate, increasing the risk of forest fires igniting (Hyde, Bohlman and Valle, 2018[288]). Power lines can also be the ignition of wildfires. For example, in California, US, power lines are estimated to cause about 10% of wildfires. Power line ignitions can occur from high winds that cause wires to touch, debris blown into power lines and wildlife interactions (nest material or animals bridging wires). At least 44 wildfires during 2014-18 were ignited by avian electrocutions in the contiguous US (i.e., excluding Hawaii and Alaska) (Barnes et al., 2022[290]). IAS in RoW can also increase fire risk (Cho, Malahlela and Ramoelo, 2015[291]). This underscores the importance of considering biodiversity in the planning, design, operation and maintenance of transmission and distribution infrastructure.

Cumulative impacts and population-level effects

Little evidence exists of the cumulative impacts of extensive transmission and distribution electricity networks on biodiversity. However, studies suggest that cumulative impacts could be significant in some ecosystems and deserve consideration. In the Amazon region, for instance, close to 40 000 km of transmission and distribution lines have been built, directly impacting approximately 23 000 km^2 of land (Hyde, Bohlman and Valle, 2018[288]). The transmission network in the Legal Amazon region is expected to be extended further, affecting intact forests and protected areas. Long-term studies of habitat fragmentation have highlighted the negative, synergistic nature of anthropogenic threats on the Amazon (Laurance et al., 2011[292]).

A study of the cumulative direct and indirect impacts of power lines in a 573 500 ha area of coastal New South Wales, Australia, concluded that the spatial footprint of power lines was <1% of landscape while the direct and indirect impacts accumulated to ~2% (10 103 ha). When also combined with roads, ~8% (33 780 ha) of habitat in the study area was potentially affected (C. Strevens, L. Puotinen and J. Whelan, 2008[293]).

The cumulative impacts from collision, electrocution and barrier effects on migratory species is also of concern. In addition to power lines, migratory birds are likely to be confronted by wind energy facilities and other human threats across their flyway. The potential for cumulative impacts on migratory species indicates the importance of international co-operation.

The evidence for adverse biodiversity impacts from the construction, operation and decommissioning of solar and wind energy facilities and power lines underscores the importance of considering biodiversity when expanding renewable power. As the boxes in this section have highlighted, impact mitigation measures exist and are being increasingly deployed (for a comprehensive overview of mitigation measures see (Bennun et al., 2021[1])). Through effective mainstreaming of biodiversity into power system planning (see Chapter 4) and appropriate policy mixes (see Chapter 5) governments can help ensure effective mitigation of these impacts and promote positive outcomes for biodiversity in renewable power projects.

3.2. Upstream impacts of renewable power infrastructure on biodiversity

Infrastructure for renewable power generation, transmission and distribution (and other low-emission technologies such as electric vehicles and batteries), place considerable demand on mineral resources. While the transition to low-emissions electricity systems will reduce coal mining, it will put increased pressure on other (critical) minerals. Low-emission technologies are becoming the fastest growing segment of mineral demand (IEA, 2021[294]), which is itself projected to increase (OECD, 2019[295]).

The extent to which the expansion of renewable power infrastructure drives up demand for minerals will depend in part on the specific technology used, material efficiency gains and the extent to which components are recycled. For example, silver and silicon metal currently dominate the solar market but are not recuperated at the end of life, unlike copper (MTE, 2020[296]).

The dependency of low-emissions infrastructure on critical minerals brings a host of economic (e.g. risk of supply chain rupture; geopolitical risks), social (e.g. human rights) and environmental challenges (IEA, 2021[294]; MTE, 2022[297]; MTE, 2020[296]; MTE, 2020[298]) (OECD, 2016[299]), one of which is mitigating the biodiversity impacts of mineral extraction and processing.

3.2.1. The material requirements of renewable power infrastructure

In a net-zero scenario by 2050, mineral demand (tonnes) from low-emission technologies increases six-fold between 2020 and 2040[7] (IEA, 2021[294]). Most of the demand is for electric vehicles (EVs) and battery storage, followed by electricity networks, solar PV and wind. The material intensity and mineral requirements of infrastructure for electricity generation vary across technologies (Månberger and Stenqvist, 2018[300]). Offshore wind has the highest material intensity (i.e., kg/MW), followed by onshore wind, and solar PV. Nuclear, coal and natural gas all have lower material intensities (Figure 3.2) (IEA, 2021[294]).

Figure 3.2. Minerals used in power generation technologies

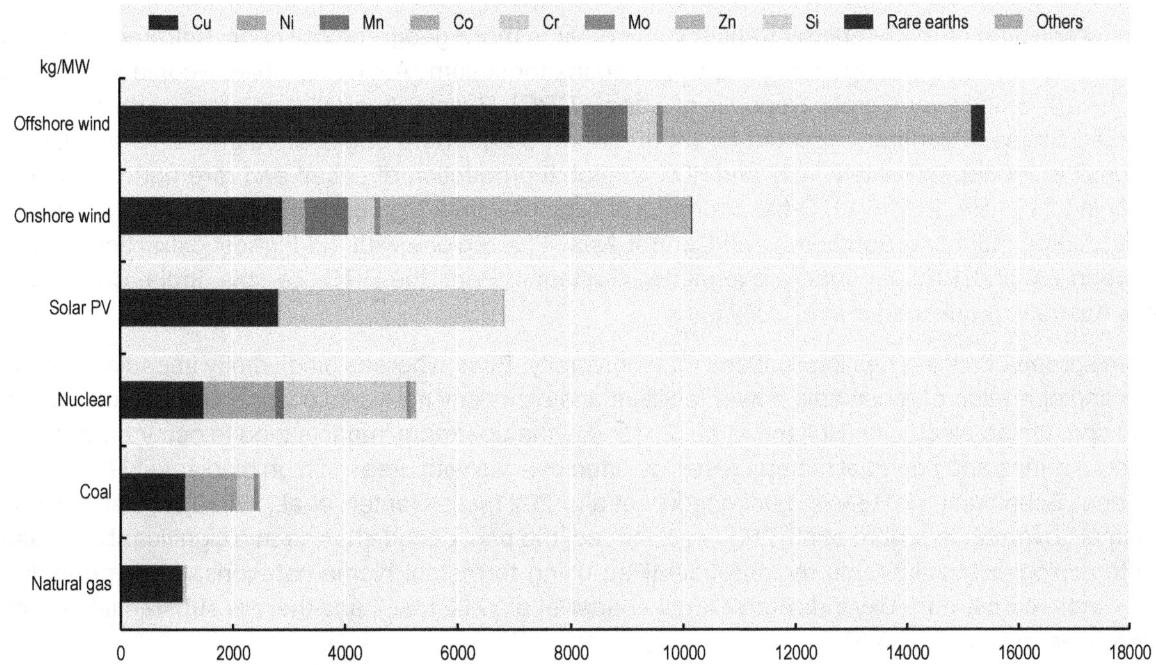

Note: Cu: Copper; Ni: Nickel; Mn: Manganese; Co: Cobalt; Cr: Chromium; Mo: Molybdenum; Zn: Zinc; Si: Silicon. Steel and aluminium are not included.
Source: (IEA, 2021[294]), The Role of Critical Minerals in Clean Energy Transitions, www.iea.org/reports/the-role-of-critical-minerals-in-clean-energy-transitions.

Wind turbines comprise various materials including concrete, steel, iron, fibreglass, polymers, aluminium, copper, zinc and rare earth elements (neodymium, praseodymium, dysprosium, and terbium). Wind turbines and low-carbon transport are the largest consumers of rare earth magnets, at 10% and 25% of the global market respectively (MTE, 2022[301]). The material intensity of turbines depends on both turbine size and type. For example, direct drive turbines have greater mineral demand than gearbox models. Permanent magnet synchronous generator models (PMSG) have greater demand on rare earth elements than double-fed induction generators and electrically excited synchronous generators (IEA, 2021[294]). Offshore turbines tend to be larger than onshore turbines, explaining their higher mineral resource demand, and are typically direct drive PMSG, therefore placing greater demand on rare earth elements.

For solar PV, key mineral resources include copper, silicon and silver. Solar PV demand for copper is projected to almost triple by 2040 in a "well below 2 °C" scenario, while demand for silicon and silver increases by 45%, with material intensity reductions reducing demand growth (IEA, 2021[294]). CSP places substantial demand on chromium, copper, manganese and nickel resources. As with wind energy, the mineral demands and intensities depend on the specific technologies used (e.g., crystalline silicon (c-Si) modules, which are the dominant PV technology, or alternatives such as cadmium telluride (CdTe), copper indium gallium diselenide (CIGS) and amorphous silicon (a-Si)).

The core components of electricity networks are iron for supporting structures (for above ground cables), copper or aluminium for wires and cables, and concrete for the foundations (RTE, 2019[302]; IEA, 2021[303]). Grid expansion to meet growing electricity demand and the integration of renewable energies could lead to a doubling in demand for copper and aluminium from electricity grids by 2040 in a "well below 2 °C" scenario (IEA, 2021[294]).

3.2.2. Geographic spread of mineral resources and overlap with biodiversity

The current production of many energy transition minerals is more geographically concentrated than that of oil or gas. For example, the top three producing nations for lithium (Australia, Chile, People's Republic of China [China]), cobalt (Democratic Republic of Congo [DRC], Russia, Australia) and rare earth elements (China, United States, Myanmar), account for more than three-quarters of global output. DRC and China were responsible for approximately 70% and 60% of global production of cobalt and rare earth elements respectively in 2019 (IEA, 2021[294]). Other countries or regions with a high concentration of mining include Brazil, West Africa, India and Southeast and Central Asia. The regions with the highest extraction growth rates (between 7% and 10% per year) are in mining clusters in Peru, the DRC, Zambia, India, China, and in Western Australia (Luckeneder et al., 2021[304]).

This uneven spread of mining has implications for biodiversity. First, whereas biodiversity impacts from the installation and operation of renewable power facilities and electricity networks occur predominantly within the country consuming electricity (Holland et al., 2019[305]), the upstream impacts tend to occur elsewhere. Second, active mines and potential mineral reserves often overlap with areas of high biodiversity (Murguía, Bringezu and Schaldach, 2016[306]; Luckeneder et al., 2021[304]; Sonter et al., 2020[307]) and weak environmental governance (Lèbre et al., 2020[308]). Indeed, the past decade has seen a significant increase of mining in ecologically vulnerable regions (identified using terrestrial biome categorisations, protected areas and water scarcity as proxy indicators) (Luckeneder et al., 2021[304]), and there is substantial debate around deep-sea mining (Box 3.5).

An analysis based on nine metals (bauxite, copper, gold, iron, lead, manganese, nickel, silver and zinc), for example, estimated that 79% of global metal ore extraction in 2019 originated from five[8] of the six most species-rich biomes[9] (Luckeneder et al., 2021[304]). In absolute values, deserts and xeric shrublands were the most exploited terrestrial biome during the past twenty years, followed by tropical and subtropical moist broadleaf forests, and temperate broadleaf and mixed forests. Mining volumes in tropical moist forest ecosystems, which are the most biodiversity-rich ecosystems on the planet, doubled from 2000-2019. Moreover, half of global metal ore extraction in 2019 took place at 20 km or less from protected areas, and 8% took place within protected areas (Luckeneder et al., 2021[304]).

Box 3.5. Deep-sea mining, biodiversity and the low-emissions transition

Deep-sea mining is a controversial activity that involves extracting mineral deposits from deep seabed (below 200 metres). Interest in deep-sea mining has arisen due to the pressure on mineral deposits on land, supply chain risks for critical minerals and the economic opportunities deep-sea mining presents. By May 2022, the International Seabed Authority (ISA) had issued 31 contracts to explore deep-sea mineral deposits and more than 1.5 million km^2 of international seabed has been set aside for mineral exploration. ISA is developing regulations to then allow deep-sea mining exploitation to begin.

However, deep-sea mining has received considerable opposition because the deep sea is poorly studied and understood. Little is known about deep-sea ecosystems and species and the impact deep-mining activities will have on them. A report commissioned by the UK Government reviewing the evidence base concluded that deep-sea mining will cause adverse impacts to the environment, affecting the composition, structure and functioning of some biological communities.

At the IUCN World Conservation Congress in Marseille in 2021, IUCN Members adopted Resolution 122 to protect deep-ocean ecosystems and species with a moratorium on deep-sea mining exploration and exploitation until certain conditions are met, including:

- Rigorous and transparent impact assessments have been conducted;
- The environmental, social, cultural and economic risks of deep seabed mining are comprehensively understood;
- The effective protection of the marine environment can be ensured;
- The precautionary principle, ecosystem approach, and the polluter pays principle have been implemented;
- Policies to ensure the responsible production and use of metals, such as the reduction of demand for primary metals, a transformation to a resource-efficient circular economy, and responsible terrestrial mining practices, have been developed and implemented; and
- Public consultation mechanisms have been incorporated into all decision-making processes related to deep-sea mining ensuring effective engagement allowing for independent review, and, where relevant, that the free, prior and informed consent of indigenous peoples is respected and consent from potentially affected communities is achieved.

Source: (IUCN Members, 2020[309]), Protection of deep-ocean ecosystems and biodiversity through a moratorium on seabed mining; (Lusty et al., 2021[310]), Deep-sea mining evidence review, www.bgs.ac.uk/download/deep-sea-mining-evidence-review/; (WWF, 2021[311]), In Too Deep: What we know, and don't know, about deep seabed mining; (Heffernan, 2019[312]), Seabed mining is coming – bringing mineral riches and fears of epic extinctions, 10.1038/d41586-019-02242-y.

3.2.3. The potential impacts of mining on biodiversity

Mining (mineral extraction and processing) can impact biodiversity in multiple ways. The extent and significance of mining impacts depend on the mineral resource and extraction method, the local species and ecosystems where the mine is found, and the socio-economic and political context (Sonter, Ali and Watson, 2018[313]).

Habitat loss and degradation are among the main direct impacts of mining. Mining can directly remove, fragment, or degrade natural habitat (Yang et al., 2013[314]; Sonter et al., 2014[315]), with the affected area ranging from <1 to several dozen km^2 in area, depending on the mineral being mined (Edwards et al., 2013[316]). This can have significant impacts on biodiversity. In Brazil, for example, mining has destroyed exceptionally diverse plant communities (Jacobi, do Carmo and de Campos, 2011[317]). Mining can also pollute air, land and water with gaseous waste (e.g. CO_2, SO_2, NO_x), particulate matter and solid wastes (Appleton et al., 2006[318]; Dudka and Adriano, 1997[319]; N Yenkie et al., 2006[320]). For example, the Indonesian Grasberg copper and gold mine, which neighbours Lorentz National Park, a World Heritage Site, has been associated with pollution of rivers and lakes in the area due to riverine tailing disposal (Martinez-Alier, 2001[321]). While many of the impacts are at a site or landscape level, mining impacts can also occur at regional or even global level (Sonter, Ali and Watson, 2018[313]). For example, sediment export from mining in the Department of *Madre de Dios* in Peru degrades ecosystems along connecting rivers in Brazil (Asner et al., 2013[322]).

The expansion of roads and railways to support mining can lead to indirect impacts on biodiversity (Edwards et al., 2013[316]), as can urban expansion around mines (Sonter et al., 2017[323]). Mining in the Amazon, for example, was found to significantly increase forest loss up to 70 km beyond mining lease boundaries, causing 11 670 km^2 of deforestation from 2005-15 (Sonter et al., 2017[323]). This deforestation is thought to result from several direct and indirect pathways including the establishment of mining infrastructure, urban expansion to support a growing workforce and the development of mineral commodity supply chains. Owing to the spatial concentration of critical minerals within landscapes and biomes, the cumulative direct and indirect impacts of multiple mining sites could be significant.

References

Adams, J. et al. (2016), *Collision and displacement vulnerability among marine birds of the California Current System associated with offshore wind energy infrastructure*, US Geological Survey, https://doi.org/10.3133/ofr20161154. [110]

Adrushchenko, Y. and V. Popenko (2012), *Birds and Power Lines in Steppe Crimea: Positive and Negative Impacts, Ukraine. – Raptors Conservation 24*, http://rrrcn.ru/en/archives/12268 (accessed on 20 July 2022). [228]

Aili, A. et al. (2022), "Radiative cooling and cold storage for concentrated solar power plants", *Energy Storage and Saving*, Vol. 1/2, pp. 93-101, https://doi.org/10.1016/j.enss.2022.02.002. [41]

Aksoy, H., F. Unal and S. Ozcan (2010), "Genotoxic Effects Of Electromagnetic Fields From High Voltage Power Lines On Some Plants", *International Journal of Environmental Rsearch*, Vol. 4/4, https://www.sid.ir/en/Journal/ViewPaper.aspx?ID=182110 (accessed on 2 August 2022). [278]

Albanito, F. et al. (2022), "Quantifying the land-based opportunity carbon costs of onshore wind farms", *Journal of Cleaner Production*, Vol. 363, p. 132480, https://doi.org/10.1016/j.jclepro.2022.132480. [198]

American Wind and Wildlife Institute (2021), *Wind Energy Interactions with Wildlife and Their Habitats: A Summary of Research Results and Priority Questions*, https://tethys.pnnl.gov/sites/default/files/publications/AWWI-Wind-Power-Wildlife-Interactions-Summary-2021.pdf. [97]

Añel, J. (ed.) (2012), "An Approach to Enhance the Conservation-Compatibility of Solar Energy Development", *PLoS ONE*, Vol. 7/6, p. e38437, https://doi.org/10.1371/journal.pone.0038437. [37]

Añel, J. (ed.) (2009), "Energy Sprawl or Energy Efficiency: Climate Policy Impacts on Natural Habitat for the United States of America", *PLoS ONE*, Vol. 4/8, p. e6802, https://doi.org/10.1371/journal.pone.0006802. [78]

Antonio Sánchez-Zapata, J. et al. (2016), “Effects of Renewable Energy Production and Infrastructure on Wildlife”, https://doi.org/10.1007/978-3-319-27912-1_5. [33]

Appleton, J. et al. (2006), “Impacts of mercury contaminated mining waste on soil quality, crops, bivalves, and fish in the Naboc River area, Mindanao, Philippines”, *Science of The Total Environment*, Vol. 354/2-3, pp. 198-211, https://doi.org/10.1016/j.scitotenv.2005.01.042. [318]

Arkumarev, V. et al. (2014), “Congregations of wintering Egyptian Vultures<i>Neophron percnopterus</i>in Afar, Ethiopia: present status and implications for conservation”, *Ostrich*, Vol. 85/2, pp. 139-145, https://doi.org/10.2989/00306525.2014.971450. [254]

Arlettaz, R. (ed.) (2011), “Minimising Mortality in Endangered Raptors Due to Power Lines: The Importance of Spatial Aggregation to Optimize the Application of Mitigation Measures”, *PLoS ONE*, Vol. 6/11, p. e28212, https://doi.org/10.1371/journal.pone.0028212. [230]

Armstrong, A., N. Ostle and J. Whitaker (2016), “Solar park microclimate and vegetation management effects on grassland carbon cycling”, *Environmental Research Letters*, Vol. 11/7, p. 074016, https://doi.org/10.1088/1748-9326/11/7/074016. [58]

Armstrong, A. et al. (2020), “Integrating environmental understanding into freshwater floatovoltaic deployment using an effects hierarchy and decision trees”, *Environmental Research Letters*, Vol. 15/11, p. 114055, https://doi.org/10.1088/1748-9326/abbf7b. [11]

Arnett, E. et al. (2008), “Patterns of Bat Fatalities at Wind Energy Facilities in North America”, *Journal of Wildlife Management*, Vol. 72/1, pp. 61-78, https://doi.org/10.2193/2007-221. [129]

Arnett, E. et al. (2010), “Altering turbine speed reduces bat mortality at wind-energy facilities”, *Frontiers in Ecology and the Environment*, Vol. 9/4, pp. 209-214, https://doi.org/10.1890/100103. [140]

Arnett, E. et al. (2009), *Effectiveness of changing wind turbine cut-in speed to reduce bat fatalities at wind facilities.*. [139]

Ashley, M., S. Mangi and L. Rodwell (2014), “The potential of offshore windfarms to act as marine protected areas – A systematic review of current evidence”, *Marine Policy*, Vol. 45, pp. 301-309, https://doi.org/10.1016/j.marpol.2013.09.002. [200]

Asner, G. et al. (2013), “Elevated rates of gold mining in the Amazon revealed through high-resolution monitoring”, *Proceedings of the National Academy of Sciences*, Vol. 110/46, pp. 18454-18459, https://doi.org/10.1073/pnas.1318271110. [322]

AWWI (2019), *Wind Turbine Interactions with Wildlife and Their Habitats A Summary of Research Results and Priority Questions Last Updated with Latest Publicly Wind Turbine Interactions with Wildlife and Their Habitats: A Summary of Research Results and Priority Questions*, https://awwi.org/resources/summary-of-wind-power-interac- (accessed on 1 July 2022). [109]

Baerwald, E. et al. (2008), “Barotrauma is a significant cause of bat fatalities at wind turbines”, *Current Biology*, Vol. 18/16, pp. R695-R696, https://doi.org/10.1016/J.CUB.2008.06.029. [90]

Baerwald, E. et al. (2009), “A Large-Scale Mitigation Experiment to Reduce Bat Fatalities at Wind Energy Facilities”, *Journal of Wildlife Management*, Vol. 73/7, pp. 1077-1081, https://doi.org/10.2193/2008-233. [141]

Bailey, H., K. Brookes and P. Thompson (2014), "Assessing environmental impacts of offshore wind farms: lessons learned and recommendations for the future", *Aquatic Biosystems*, Vol. 10/1, p. 8, https://doi.org/10.1186/2046-9063-10-8. [183]

Barber, C. et al. (2014), "Roads, deforestation, and the mitigating effect of protected areas in the Amazon", *Biological Conservation*, Vol. 177, pp. 203-209, https://doi.org/10.1016/j.biocon.2014.07.004. [286]

Barnes, T. et al. (2022), "Wildland fires ignited by avian electrocutions", *Wildlife Society Bulletin*, https://doi.org/10.1002/wsb.1302. [290]

Barré, K. et al. (2018), "Estimating habitat loss due to wind turbine avoidance by bats: Implications for European siting guidance", *Biological Conservation*, Vol. 226, pp. 205-214, https://doi.org/10.1016/j.biocon.2018.07.011. [171]

Barrios, L. and A. Rodríguez (2004), "Behavioural and environmental correlates of soaring-bird mortality at on-shore wind turbines", *Journal of Applied Ecology*, Vol. 41/1, pp. 72-81, https://doi.org/10.1111/j.1365-2664.2004.00876.x. [114]

Barron-Gafford, G. et al. (2016), "The Photovoltaic Heat Island Effect: Larger solar power plants increase local temperatures", *Scientific Reports*, Vol. 6/1, https://doi.org/10.1038/srep35070. [54]

Bartzke, G. et al. (2014), "International Journal of Biodiversity and Conservation The effects of power lines on ungulates and implications for power line routing and rights-of-way management", Vol. 6/9, pp. 647-662, https://doi.org/10.5897/IJBC2014.0716. [261]

Bennun, L. et al. (2021), *Mitigating biodiversity impacts associated with solar and wind energy development: guidelines for project developers*, IUCN, International Union for Conservation of Nature, https://doi.org/10.2305/iucn.ch.2021.04.en. [1]

Berger-Tal, O. and D. Saltz (2019), "Invisible barriers: anthropogenic impacts on inter- and intra-specific interactions as drivers of landscape-independent fragmentation", *Philosophical Transactions of the Royal Society B: Biological Sciences*, Vol. 374/1781, p. 20180049, https://doi.org/10.1098/rstb.2018.0049. [45]

Bergström, L., F. Sundqvist and U. Bergström (2013), "Effects of an offshore wind farm on temporal and spatial patterns in the demersal fish community", *Marine Ecology Progress Series*, Vol. 485, pp. 199-210, https://doi.org/10.3354/meps10344. [160]

Bernardino, J. et al. (2018), "Bird collisions with power lines: State of the art and priority areas for research", *Biological Conservation*, Vol. 222, pp. 1-13, https://doi.org/10.1016/j.biocon.2018.02.029. [220]

Bevk, T. and M. Golobič (2020), "Contentious eye-catchers: Perceptions of landscapes changed by solar power plants in Slovenia", *Renewable Energy*, Vol. 152, pp. 999-1010, https://doi.org/10.1016/j.renene.2020.01.108. [70]

Biasotto, L. and A. Kindel (2018), "Power lines and impacts on biodiversity: A systematic review", *Environmental Impact Assessment Review*, Vol. 71, pp. 110-119, https://doi.org/10.1016/j.eiar.2018.04.010. [212]

Bijleveld, A. (ed.) (2019), "Wind turbines cause functional habitat loss for migratory soaring birds", *Journal of Animal Ecology*, Vol. 89/1, pp. 93-103, https://doi.org/10.1111/1365-2656.12961. [170]

Biogradlija, A. (2022), *RWE looks into black rotor wind blades for birds protection*, https://www.industryandenergy.eu/renewables/rwe-looks-into-black-rotor-wind-blades-for-birds-protection/ (accessed on 11 October 2022). [142]

Boinski, S. et al. (1998), "Squirrel monkeys in Costa Rica: drifting to extinction", *Oryx*, Vol. 32/01, p. 45, https://doi.org/10.1017/s0030605300029756. [239]

Brand, L. et al. (2016), "Mitigation-driven translocation effects on temperature, condition, growth, and mortality of Mojave desert tortoise (Gopherus agassizii) in the face of solar energy development", *Biological Conservation*, Vol. 200, pp. 104-111, https://doi.org/10.1016/j.biocon.2016.05.032. [63]

Brandt, M. et al. (2018), "Disturbance of harbour porpoises during construction of the first seven offshore wind farms in Germany", *Marine Ecology Progress Series*, Vol. 596, pp. 213-232, https://doi.org/10.3354/meps12560. [179]

Brigham, R. (ed.) (2014), "A Comprehensive Analysis of Small-Passerine Fatalities from Collision with Turbines at Wind Energy Facilities", *PLoS ONE*, Vol. 9/9, p. e107491, https://doi.org/10.1371/journal.pone.0107491. [95]

Brignon, J. et al. (2022), "A risk-based method to prioritize cumulative impacts assessment on marine biodiversity and research policy for offshore wind farms in France", *Environmental Science & Policy*, Vol. 128, pp. 264-276, https://doi.org/10.1016/j.envsci.2021.12.003. [210]

Brown, T. (2015), "Transmission network loading in Europe with high shares of renewables", *IET Renewable Power Generation*, Vol. 9/1, pp. 57-65, https://doi.org/10.1049/iet-rpg.2014.0114. [215]

Brunel (2021), *What is Concentrated Solar Power and how does CSP work? - Brunel*, https://www.brunel.net/en/blog/renewable-energy/concentrated-solar-power (accessed on 1 March 2022). [16]

Burda, H. et al. (2009), "Extremely low-frequency electromagnetic fields disrupt magnetic alignment of ruminants", *Proceedings of the National Academy of Sciences*, Vol. 106/14, pp. 5708-5713, https://doi.org/10.1073/pnas.0811194106. [277]

Buton, C. (2023), *Impacts écologiques des clôtures et solutions de remédiation possibles.*, https://www.trameverteetbleue.fr/sites/default/files/references_bibliographiques/impacts_ecologiques_des_clotures_bp_cpv_2023-07-28.pdf. [52]

C. Strevens, T., M. L. Puotinen and R. J. Whelan (2008), "Powerline easements: ecological impacts and contribution to habitat fragmentation from linear features", *Pacific Conservation Biology*, Vol. 14/3, p. 159, https://doi.org/10.1071/pc080159. [293]

Cabrera-Cruz, S. et al. (2020), "Estimates of aerial vertebrate mortality at wind farms in a bird migration corridor and bat diversity hotspot", *Global Ecology and Conservation*, Vol. 22, p. e00966, https://doi.org/10.1016/j.gecco.2020.e00966. [101]

Campbell, M. et al. (2009), "Minimizing utility-scale PV power plant LCOE through the use of high capacity factor configurations", *2009 34th IEEE Photovoltaic Specialists Conference (PVSC)*, https://doi.org/10.1109/pvsc.2009.5411650. [6]

Castillo, S. (2021), *Desert tortoise deaths raise questions as Nevada solar farms grow | Las Vegas Review-Journal*, https://www.reviewjournal.com/news/science-and-technology/desert-tortoise-deaths-raise-concerns-as-solar-farms-solve-energy-need-2408456/ (accessed on 26 July 2022). [64]

Cecala, K., W. Lowe and J. Maerz (2014), "Riparian disturbance restricts in-stream movement of salamanders", *Freshwater Biology*, Vol. 59/11, pp. 2354-2364, https://doi.org/10.1111/fwb.12439. [253]

Chapman, M. (ed.) (2015), "Urban Power Line Corridors as Novel Habitats for Grassland and Alien Plant Species in South-Western Finland", *PLOS ONE*, Vol. 10/11, p. e0142236, https://doi.org/10.1371/journal.pone.0142236. [284]

Chock, R. et al. (2020), "Evaluating potential effects of solar power facilities on wildlife from an animal behavior perspective", *Conservation Science and Practice*, Vol. 3/2, https://doi.org/10.1111/csp2.319. [30]

Choi, Y. (2014), "A Study on Power Generation Analysis of Floating PV System Considering Environmental Impact", *International Journal of Software Engineering and Its Applications*, Vol. 8/1, pp. 75-84, https://doi.org/10.14257/ijseia.2014.8.1.07. [14]

Cho, M., O. Malahlela and A. Ramoelo (2015), "Assessing the utility WorldView-2 imagery for tree species mapping in South African subtropical humid forest and the conservation implications: Dukuduku forest patch as case study", *International Journal of Applied Earth Observation and Geoinformation*, Vol. 38, pp. 349-357, https://doi.org/10.1016/j.jag.2015.01.015. [291]

Clarke, D., K. Pearce and J. White (2006), "Powerline corridors: degraded ecosystems or wildlife havens?", *Wildlife Research*, Vol. 33/8, p. 615, https://doi.org/10.1071/wr05085. [260]

Colman, J. et al. (2015), "High-voltage power lines near wild reindeer calving areas", *European Journal of Wildlife Research*, Vol. 61/6, pp. 881-893, https://doi.org/10.1007/s10344-015-0965-x. [273]

Conkling, T. et al. (2022), "Vulnerability of avian populations to renewable energy production", *Royal Society Open Science*, Vol. 9/3, https://doi.org/10.1098/rsos.211558. [80]

Copping, A. and L. Hemery (2020), *OES-Environmental 2020 State of the Science Report: Environmental Effects of Marine Renewable Energy Development Around the World. Report for Ocean Energy Systems (OES)*, Office of Scientific and Technical Information (OSTI), https://doi.org/10.2172/1632878. [280]

Costello, K. et al. (2022), "Assessing the potential for invasive species introductions and secondary spread using vessel movements in maritime ports", *Marine Pollution Bulletin*, Vol. 177, p. 113496, https://doi.org/10.1016/j.marpolbul.2022.113496. [191]

Cryan, P. (2014), "Behavior of bats at wind turbines", *Proc. Natl. Acad. Sci.*, Vol. 111/42, pp. 15126-15131, https://doi.org/10.1073/pnas.1406672111. [121]

Cryan, P. (2008), "Mating Behavior as a Possible Cause of Bat Fatalities at Wind Turbines", *Journal of Wildlife Management*, Vol. 72/3, pp. 845-849, https://doi.org/10.2193/2007-371. [122]

Cypher, B. et al. (2021), "Photovoltaic solar farms in California: can we have renewable electricity and our species, too?", *California Fish and Wildlife Journal*, Vol. 107/3, pp. 231-248, https://doi.org/10.51492/cfwj.hwisi.6. [53]

Dähne, M. et al. (2013), "Effects of pile-driving on harbour porpoises (Phocoena phocoena) at the first offshore wind farm in Germany", *Environmental Research Letters*, Vol. 8/2, p. 025002, https://doi.org/10.1088/1748-9326/8/2/025002. [178]

Dai, K. et al. (2015), "Environmental issues associated with wind energy - A review", *Renewable Energy*, Vol. 75, pp. 911-921, https://doi.org/10.1016/J.RENENE.2014.10.074. [151]

de Lucas, M. et al. (2008), "Collision fatality of raptors in wind farms does not depend on raptor abundance", *Journal of Applied Ecology*, Vol. 45/6, pp. 1695-1703, https://doi.org/10.1111/j.1365-2664.2008.01549.x. [115]

De Marco, A. et al. (2014), "The contribution of Utility-Scale Solar Energy to the global climate regulation and its effects on local ecosystem services", *Global Ecology and Conservation*, Vol. 2, pp. 324-337, https://doi.org/10.1016/J.GECCO.2014.10.010. [72]

De Mesel, I. et al. (2015), "Succession and seasonal dynamics of the epifauna community on offshore wind farm foundations and their role as stepping stones for non-indigenous species", *Hydrobiologia*, Vol. 756/1, pp. 37-50, https://doi.org/10.1007/s10750-014-2157-1. [159]

Dhar, A. et al. (2020), "Perspectives on environmental impacts and a land reclamation strategy for solar and wind energy systems", *Science of The Total Environment*, Vol. 718, p. 134602, https://doi.org/10.1016/j.scitotenv.2019.134602. [150]

Didderen, K. (2019), *Pilot to actively restore native oyster reefs in theNorth Sea: comprehensive report to share lessons learned in 2018*. [162]

Diffendorfer, J. et al. (2019), "Geographic context affects the landscape change and fragmentation caused by wind energy facilities", *PeerJ*, Vol. 7, p. e7129, https://doi.org/10.7717/peerj.7129. [155]

Diffendorfer, J. et al. (2021), "Demographic and potential biological removal models identify raptor species sensitive to current and future wind energy", *Ecosphere*, Vol. 12/6, https://doi.org/10.1002/ecs2.3531. [209]

Ding, H. (ed.) (2020), "An investigation into the potential for wind turbines to cause barotrauma in bats", *PLOS ONE*, Vol. 15/12, p. e0242485, https://doi.org/10.1371/journal.pone.0242485. [92]

Dixon, A. et al. (2018), *Efficacy of a mitigation method to reduce raptor electrocution at an electricity distribution*, Conservation Evidence, https://conservationevidencejournal.com/reference/pdf/6861. [242]

Dixon, A. et al. (2013), "The Use of Power Lines by Breeding Raptors and Corvids in Mongolia: Nest-Site Characteristics and Management Using Artificial Nests", *Journal of Raptor Research*, Vol. 47/3, pp. 282-291, https://doi.org/10.3356/jrr-12-00020.1. [256]

DoE (2023), *U.S. Installed and Potential Wind Capacity Generation*, https://windexchange.energy.gov/maps-data/321. [98]

Dörenkämper, M. et al. (2021), “The cooling effect of floating PV in two different climate zones: A comparison of field test data from the Netherlands and Singapore”, *Solar Energy*, Vol. 214, pp. 239-247, https://doi.org/10.1016/j.solener.2020.11.029. [12]

Drewitt, A. and R. Langston (2008), “Collision Effects of Wind-power Generators and Other Obstacles on Birds”, *Annals of the New York Academy of Sciences*, Vol. 1134/1, pp. 233-266, https://doi.org/10.1196/annals.1439.015. [125]

Dröes, M. and H. Koster (2021), “Wind turbines, solar farms, and house prices”, *Energy Policy*, Vol. 155, p. 112327, https://doi.org/10.1016/j.enpol.2021.112327. [195]

Dubé, C., S. Pellerin and M. Poulin (2011), “Do power line rights-of-way facilitate the spread of non-peatland and invasive plants in bogs and fens?”, *Botany*, Vol. 89/2, pp. 91-103, https://doi.org/10.1139/b10-089. [283]

Dudka, S. and D. Adriano (1997), “Environmental Impacts of Metal Ore Mining and Processing: A Review”, *Journal of Environmental Quality*, Vol. 26/3, pp. 590-602, https://doi.org/10.2134/jeq1997.00472425002600030003x. [319]

Dunlop, J. (2010), *Photovoltaic Systems. 3rd Edition, American Technical Publishers, Inc., Orland Park. - References - Scientific Research Publishing*, American Technical Publishers, Inc. Orland Park, https://www.scirp.org/(S(lz5mqp453ed%20snp55rrgjct55))/reference/referencespapers.aspx?referenceid=1697608 (accessed on 14 June 2022). [7]

EC (2021), *OFFSHORE WIND AND FISHERIES | The European Maritime Spatial Planning Platform*, https://maritime-spatial-planning.ec.europa.eu/sector-information/offshore-wind-and-fisheries (accessed on 19 July 2022). [204]

EC (2019), *Water consumption solution for efficienct concentrated solar power*, https://ec.europa.eu/research-and-innovation/en/projects/success-stories/all/water-consumption-solution-efficient-concentrated-solar-power#:~:text=As%20a%20result%2C%20they%20can,natural%20gas%2Dfired%20power%20plants. [42]

EC (2019), *Water consumption solution for efficient concentrated solar power*, https://ec.europa.eu/research-and-innovation/en/projects/success-stories/all/water-consumption-solution-efficient-concentrated-solar-power. [39]

Edwards, D. et al. (2013), “Mining and the African Environment”, *Conservation Letters*, Vol. 7/3, pp. 302-311, https://doi.org/10.1111/conl.12076. [316]

EU (2019), *Clean energy for all Europeans package*, https://energy.ec.europa.eu/topics/energy-strategy/clean-energy-all-europeans-package_en (accessed on 20 July 2022). [218]

Europacable (2021), *ELECTRICITY TRANSMISSION OF TOMORROW UNDERGROUND AND SUBSEA CABLES IN EUROPE ®*, https://europacable.eu/wp-content/uploads/2021/01/Europacable-Brochure-FINAL_Web-File.pdf (accessed on 20 July 2022). [219]

European Commission (2020), *Commission notice Guidance document on wind energy developments and EU nature legislation Commission notice Guidance document on wind energy developments and EU nature legislation Guidance document on wind energy developments and EU Nature Legislation*, European Commission (C2020) 7730 final, Brussels, https://ec.europa.eu/environment/nature/natura2000/management/docs/wind_farms_en.pdf (accessed on 3 May 2022). [89]

Everaert, J. and E. Stienen (2006), "Impact of wind turbines on birds in Zeebrugge (Belgium)", *Biodiversity and Conservation*, Vol. 16/12, pp. 3345-3359, https://doi.org/10.1007/s10531-006-9082-1. [111]

Everaert, J. and E. Stienen (2006), "Impact of wind turbines on birds in Zeebrugge (Belgium)", in *Biodiversity and Conservation in Europe, Topics in Biodiversity and Conservation*, Springer Netherlands, Dordrecht, https://doi.org/10.1007/978-1-4020-6865-2_8. [126]

Exley, G. et al. (2021), "Scientific and stakeholder evidence-based assessment: Ecosystem response to floating solar photovoltaics and implications for sustainability", *Renewable and Sustainable Energy Reviews*, Vol. 152, https://doi.org/10.1016/J.RSER.2021.111639. [18]

Exley, G. et al. (2021), "Scientific and stakeholder evidence-based assessment: Ecosystem response to floating solar photovoltaics and implications for sustainability", *Renewable and Sustainable Energy Reviews*, Vol. 152, p. 111639, https://doi.org/10.1016/j.rser.2021.111639. [68]

Fahrig, L. (2003), "Effects of Habitat Fragmentation on Biodiversity", *Annual Review of Ecology, Evolution, and Systematics*, Vol. 34/1, pp. 487-515, https://doi.org/10.1146/annurev.ecolsys.34.011802.132419. [47]

Farr, H. et al. (2021), "Potential environmental effects of deepwater floating offshore wind energy facilities", *Ocean & Coastal Management*, Vol. 207, p. 105611, https://doi.org/10.1016/j.ocecoaman.2021.105611. [274]

Fernández-Bellon, D. et al. (2018), "Effects of development of wind energy and associated changes in land use on bird densities in upland areas", *Conservation Biology*, Vol. 33/2, pp. 413-422, https://doi.org/10.1111/cobi.13239. [165]

Fernie, K. and S. Reynolds (2005), "The Effects of Electromagnetic Fields From Power Lines on Avian Reproductive Biology and Physiology: A Review", *Journal of Toxicology and Environmental Health, Part B*, Vol. 8/2, pp. 127-140, https://doi.org/10.1080/10937400590909022. [275]

Ferrão da Costa, G. et al. (2017), "The Indirect Impacts of Wind Farms on Terrestrial Mammals: Insights from the Disturbance and Exclusion Effects on Wolves (Canis lupus)", in *Biodiversity and Wind Farms in Portugal*, Springer International Publishing, Cham, https://doi.org/10.1007/978-3-319-60351-3_5. [186]

Ferrer, M. et al. (2022), "Significant decline of Griffon Vulture collision mortality in wind farms during 13-year of a selective turbine stopping protocol", *Global Ecology and Conservation*, Vol. 38, p. e02203, https://doi.org/10.1016/j.gecco.2022.e02203. [143]

Frederiksen, M. (ed.) (2014), "Offshore marine renewable energy devices as stepping stones across biogeographical boundaries", *Journal of Applied Ecology*, Vol. 51/2, pp. 330-338, https://doi.org/10.1111/1365-2664.12207. [193]

Frick, W. et al. (2017), "Fatalities at wind turbines may threaten population viability of a migratory bat", *Biological Conservation*, Vol. 209, pp. 172-177, https://doi.org/10.1016/j.biocon.2017.02.023. [205]

Frick, W., T. Kingston and J. Flanders (2019), "A review of the major threats and challenges to global bat conservation", *Annals of the New York Academy of Sciences*, Vol. 1469/1, pp. 5-25, https://doi.org/10.1111/nyas.14045. [208]

Fthenakis, V. and H. Kim (2009), "Land use and electricity generation: A life-cycle analysis", *Renewable and Sustainable Energy Reviews*, Vol. 13/6-7, pp. 1465-1474, https://doi.org/10.1016/j.rser.2008.09.017. [20]

Germano, J. et al. (2015), "Mitigation-driven translocations: are we moving wildlife in the right direction?", *Frontiers in Ecology and the Environment*, Vol. 13/2, pp. 100-105, https://doi.org/10.1890/140137. [62]

Gibbons, S. (2015), "Gone with the wind: Valuing the visual impacts of wind turbines through house prices", *Journal of Environmental Economics and Management*, Vol. 72, pp. 177-196, https://doi.org/10.1016/j.jeem.2015.04.006. [194]

Gibson, D. et al. (2018), "Effects of power lines on habitat use and demography of greater sage-grouse (<i>Centrocercus urophasianus</i>)", *Wildlife Monographs*, Vol. 200/1, pp. 1-41, https://doi.org/10.1002/wmon.1034. [270]

Gill, A. et al. (2014), "Marine Renewable Energy, Electromagnetic (EM) Fields and EM-Sensitive Animals", in *Marine Renewable Energy Technology and Environmental Interactions, Humanity and the Sea*, Springer Netherlands, Dordrecht, https://doi.org/10.1007/978-94-017-8002-5_6. [281]

Gillan, J. et al. (2013), "Using spatial statistics and point-pattern simulations to assess the spatial dependency between greater sage-grouse and anthropogenic features", *Wildlife Society Bulletin*, Vol. 37/2, pp. 301-310, https://doi.org/10.1002/wsb.272. [266]

Goodale, M. and A. Milman (2019), "Assessing the cumulative exposure of wildlife to offshore wind energy development", *Journal of Environmental Management*, Vol. 235, pp. 77-83, https://doi.org/10.1016/j.jenvman.2019.01.022. [5]

Goosem, M. and H. Marsh (1997), "Fragmentation of a Small-mammal Community by a Powerline Corridor through Tropical Rainforest", *Wildlife Research*, Vol. 24/5, p. 613, https://doi.org/10.1071/wr96063. [251]

Graham, I. et al. (2019), "Harbour porpoise responses to pile-driving diminish over time", *Royal Society Open Science*, Vol. 6/6, p. 190335, https://doi.org/10.1098/rsos.190335. [180]

Grippo, M., J. Hayse and B. O'Connor (2014), "Solar Energy Development and Aquatic Ecosystems in the Southwestern United States: Potential Impacts, Mitigation, and Research Needs", *Environmental Management*, Vol. 55/1, pp. 244-256, https://doi.org/10.1007/s00267-014-0384-x. [38]

Grodsky, S., J. Campbell and R. Hernandez (2021), "Solar energy development impacts flower-visiting beetles and flies in the Mojave Desert", *Biological Conservation*, Vol. 263, p. 109336, https://doi.org/10.1016/j.biocon.2021.109336. [73]

Grodsky, S. and R. Hernandez (2020), "Reduced ecosystem services of desert plants from ground-mounted solar energy development", *Nature Sustainability*, Vol. 3/12, pp. 1036-1043, https://doi.org/10.1038/s41893-020-0574-x. [35]

Guerin, T. (2017), "A case study identifying and mitigating the environmental and community impacts from construction of a utility-scale solar photovoltaic power plant in eastern Australia", *Solar Energy*, Vol. 146, pp. 94-104, https://doi.org/10.1016/J.SOLENER.2017.02.020. [50]

Guil, F. and J. Pérez-García (2022), "Bird electrocution on power lines: Spatial gaps and identification of driving factors at global scales", *Journal of Environmental Management*, Vol. 301, p. 113890, https://doi.org/10.1016/j.jenvman.2021.113890. [240]

GWEC (2022), *Global Wind Report 2022*, https://gwec.net/global-wind-report-2022/. [83]

Haas, D. et al. (2005), "Protecting birds on powerlines: a practical guide on the risks to birds from electricity transmission facilities and how to minimise any such adverse effects", http://birdsandpowerlines.org/cm/media/Protecting_birds_on_powerlines.pdf (accessed on 20 July 2022). [243]

Haddad, N. et al. (2015), "Habitat fragmentation and its lasting impact on Earth's ecosystems", *Science Advances*, Vol. 1/2, https://doi.org/10.1126/sciadv.1500052. [46]

Hallmann, C. et al. (2017), "More than 75 percent decline over 27 years in total flying insect biomass in protected areas", https://doi.org/10.1371/journal.pone.0185809. [133]

Harsanyi, P. et al. (2022), "The Effects of Anthropogenic Electromagnetic Fields (EMF) on the Early Development of Two Commercially Important Crustaceans, European Lobster, Homarus gammarus (L.) and Edible Crab, Cancer pagurus (L.)", *Journal of Marine Science and Engineering*, Vol. 10/5, p. 564, https://doi.org/10.3390/jmse10050564. [282]

Hastik, R. et al. (2015), "Renewable energies and ecosystem service impacts", *Renewable and Sustainable Energy Reviews*, Vol. 48, pp. 608-623, https://doi.org/10.1016/j.rser.2015.04.004. [71]

Hayes, M. (2013), "Bats Killed in Large Numbers at United States Wind Energy Facilities", *BioScience*, Vol. 63/12, pp. 975-979, https://doi.org/10.1525/bio.2013.63.12.10. [104]

Heffernan, O. (2019), "Seabed mining is coming — bringing mineral riches and fears of epic extinctions", *Nature*, Vol. 571/7766, pp. 465-468, https://doi.org/10.1038/d41586-019-02242-y. [312]

Hermy, M. (ed.) (2013), "Edge effects of linear canopy openings on understorey communities in a lowland Atlantic tropical forest", *Applied Vegetation Science*, Vol. 17/1, pp. 121-128, https://doi.org/10.1111/avsc.12043. [249]

Hernández-Matías, A. et al. (2015), "Electrocution threatens the viability of populations of the endangered Bonelli's eagle (Aquila fasciata) in Southern Europe", *Biological Conservation*, Vol. 191, pp. 110-116, https://doi.org/10.1016/j.biocon.2015.06.028. [234]

Hernandez, R. et al. (2014), "Environmental impacts of utility-scale solar energy", *Renewable and Sustainable Energy Reviews*, Vol. 29, pp. 766-779, https://doi.org/10.1016/j.rser.2013.08.041. [9]

Hernandez, R., M. Hoffacker and C. Field (2014), "Land-Use Efficiency of Big Solar", *Environmental Science & Technology*, Vol. 48/2, pp. 1315-1323, https://doi.org/10.1021/es4043726. [77]

Hernandez, R. et al. (2015), "Solar energy development impacts on land cover change and protected areas", *Proceedings of the National Academy of Sciences*, Vol. 112/44, pp. 13579-13584, https://doi.org/10.1073/PNAS.1517656112. [48]

Heuck, C. et al. (2019), "Wind turbines in high quality habitat cause disproportionate increases in collision mortality of the white-tailed eagle", *Biological Conservation*, Vol. 236, pp. 44-51, https://doi.org/10.1016/j.biocon.2019.05.018. [127]

Ho, C. (2016), "Review of avian mortality studies at concentrating solar power plants", https://doi.org/10.1063/1.4949164. [28]

Holland, R. et al. (2019), "The influence of the global electric power system on terrestrial biodiversity", *Proceedings of the National Academy of Sciences*, Vol. 116/51, pp. 26078-26084, https://doi.org/10.1073/pnas.1909269116. [305]

Hooftman, D. (ed.) (2015), "Edge effects on plant communities along power line clearings", *Journal of Applied Ecology*, Vol. 52/4, pp. 871-880, https://doi.org/10.1111/1365-2664.12460. [263]

Hooper, T., A. Armstrong and B. Vlaswinkel (2021), "Environmental impacts and benefits of marine floating solar", *Solar Energy*, Vol. 219, pp. 11-14, https://doi.org/10.1016/j.solener.2020.10.010. [10]

Horn, J., E. Arnett and T. Kunz (2008), "Behavioral Responses of Bats to Operating Wind Turbines", *Journal of Wildlife Management*, Vol. 72/1, pp. 123-132, https://doi.org/10.2193/2006-465. [144]

Horváth, G. (2010), "Reducing the Maladaptive Attractiveness of Solar Panels to Polarotactic Insects", *Conservation Biology*, Vol. 24/6, pp. 1644-1653, https://doi.org/10.1111/j.1523-1739.2010.01518.x. [66]

Horváth, G. et al. (2009), "Polarized light pollution: a new kind of ecological photopollution", *Frontiers in Ecology and the Environment*, Vol. 7/6, pp. 317-325, https://doi.org/10.1890/080129. [65]

Howe, K., P. Coates and D. Delehanty (2014), "Selection of anthropogenic features and vegetation characteristics by nesting Common Ravens in the sagebrush ecosystem", *The Condor*, Vol. 116/1, pp. 35-49, https://doi.org/10.1650/condor-13-115-r2.1. [257]

Hüppop, O. (2019), *Migratory birds and bats.*, Pelagic Publishing. [93]

Huso, M. et al. (2021), "Relative energy production determines effect of repowering on wildlife mortality at wind energy facilities", *Journal of Applied Ecology*, Vol. 58/6, pp. 1284-1290, https://doi.org/10.1111/1365-2664.13853. [145]

Huso, M., T. Dietsch and C. Nicolai (2016), *Mortality Monitoring Design for Utility-Scale Solar Power Facilities*, https://doi.org/10.3133/ofr20161087 (accessed on 22 February 2022). [29]

Hyde, J., S. Bohlman and D. Valle (2018), "Transmission lines are an under-acknowledged conservation threat to the Brazilian Amazon", *Biological Conservation*, Vol. 228, pp. 343-356, https://doi.org/10.1016/j.biocon.2018.10.027. [288]

IBRD (2020), *Concentrating Solar Power Clean Power on Demand 24/7*, http://www.worldbank.org (accessed on 15 December 2021). [17]

IEA (2021), *The Role of Critical Minerals in Clean Energy Transitions*, IEA, https://www.iea.org/reports/the-role-of-critical-minerals-in-clean-energy-transitions. [303]

IEA (2021), *Net Zero by 2050: A Roadmap for the Global Energy Sector*, OECD Publishing, Paris, https://doi.org/10.1787/c8328405-en. [216]

IEA (2021), *The Role of Critical Minerals in Clean Energy Transitions*, International Energy Agency, https://www.iea.org/reports/the-role-of-critical-minerals-in-clean-energy-transitions. [294]

IEA (2020), *Renewables 2020.* [81]

IEA (2019), *Offshore Wind Outlook*, https://iea.blob.core.windows.net/assets/495ab264-4ddf-4b68-b9c0-514295ff40a7/Offshore_Wind_Outlook_2019.pdf. [84]

IEA ETSAP (2014), *Electricity Transmission and Distribution*, https://iea-etsap.org/E-TechDS/PDF/E12_el-t&d_KV_Apr2014_GSOK.pdf (accessed on 20 July 2022). [217]

IFC (2013), *Good Practice Handbook on Cumulative Impact Assessment and Management: Guidance for the Private Sector in Emerging Markets*, https://www.ifc.org/wps/wcm/connect/topics_ext_content/ifc_external_corporate_site/sustainability-at-ifc/publications/publications_handbook_cumulativeimpactassessment (accessed on 1 July 2022). [4]

Inger, R. et al. (2009), "Marine renewable energy: potential benefits to biodiversity? An urgent call for research", *Journal of Applied Ecology*, https://doi.org/10.1111/j.1365-2664.2009.01697.x. [135]

IPCC (2011), *Special Report on Renewable Energy Sources and Climate Change Mitigation.*. [85]

IRENA (2019), *Future of wind - Deployment, investment, technology, grid integration and socioeconomic aspects*. [88]

IUCN Members (2020), *Protection of deep-ocean ecosystems and biodiversity through a moratorium on seabed mining*, https://portals.iucn.org/library/sites/library/files/resrecfiles/WCC_2020_RES_122_EN.pdf. [309]

Jacobi, C., F. do Carmo and I. de Campos (2011), "Soaring Extinction Threats to Endemic Plants in Brazilian Metal-Rich Regions", *AMBIO*, Vol. 40/5, pp. 540-543, https://doi.org/10.1007/s13280-011-0151-7. [317]

Jeal, C. et al. (2019), "Impacts of a concentrated solar power trough facility on birds and other wildlife in South Africa", *Ostrich*, Vol. 90/2, pp. 129-137, https://doi.org/10.2989/00306525.2019.1581296. [22]

Jenkins, A., J. Smallie and M. Diamond (2010), "Avian collisions with power lines: a global review of causes and mitigation with a South African perspective", *Bird Conservation International*, Vol. 20/3, pp. 263-278, https://doi.org/10.1017/s0959270910000122. [244]

JNCC (2010), *Statutory nature conservation agency protocol for minimising the risk of injury to marine mammals from piling noise*, https://data.jncc.gov.uk/data/31662b6a-19ed-4918-9fab-8fbcff752046/JNCC-CNCB-Piling-protocol-August2010-Web.pdf. [190]

Kagan, R. et al. (2014), *Avian Mortality at Solar Energy Facilities in Southern California: A Preliminary Analysis*, National Fish and Wildlife Forensics Laboratory, https://www.ourenergypolicy.org/wp-content/uploads/2014/04/avian-mortality.pdf (accessed on 20 December 2021). [25]

Kasso, M. and M. Balakrishnan (2013), "Ecological and Economic Importance of Bats (Order Chiroptera)", *ISRN Biodiversity*, Vol. 2013, pp. 1-9, https://doi.org/10.1155/2013/187415. [211]

Katsis, L. et al. (2018), "Spatial Patterns of Primate Electrocutions in Diani, Kenya", *International Journal of Primatology*, Vol. 39/4, pp. 493-510, https://doi.org/10.1007/s10764-018-0046-6. [235]

Keehn, J. and C. Feldman (2018), "Disturbance affects biotic community composition at desert wind farms", *Wildlife Research*, Vol. 45/5, p. 383, https://doi.org/10.1071/wr17059. [201]

Kim, J. et al. (2021), "Current site planning of medium to large solar power systems accelerates the loss of the remaining semi-natural and agricultural habitats", *Science of The Total Environment*, Vol. 779, p. 146475, https://doi.org/10.1016/j.scitotenv.2021.146475. [79]

King, D. et al. (2009), "Effects of width, edge and habitat on the abundance and nesting success of scrub–shrub birds in powerline corridors", *Biological Conservation*, Vol. 142/11, pp. 2672-2680, https://doi.org/10.1016/j.biocon.2009.06.016. [255]

Klich, D. et al. (2020), "Roe deer stress response to a wind farms: Methodological and practical implications", *Ecological Indicators*, Vol. 117, p. 106658, https://doi.org/10.1016/j.ecolind.2020.106658. [185]

Kok, A. et al. (2021), "An echosounder view on the potential effects of impulsive noise pollution on pelagic fish around windfarms in the North Sea", *Environmental Pollution*, Vol. 290, p. 118063, https://doi.org/10.1016/j.envpol.2021.118063. [182]

Kolnegari et al. (2020), "Electrical Components Involved in Avian-Caused Outages in Iran", *Bird Conservation International*, Vol. 31/3, pp. 364-378, https://doi.org/10.1017/s0959270920000507. [233]

Kosciuch, K. et al. (2021), "Aquatic Habitat Bird Occurrences at Photovoltaic Solar Energy Development in Southern California, USA", *Diversity*, Vol. 13/11, p. 524, https://doi.org/10.3390/d13110524. [67]

Krüger, S., R. Simmons and A. Amar (2015), "Anthropogenic activities influence the abandonment of Bearded Vulture (<i>Gypaetus barbatus</i>) territories in southern Africa", *The Condor*, Vol. 117/1, pp. 94-107, https://doi.org/10.1650/condor-14-121.1. [267]

Kumara, H. et al. (2022), "Responses of birds and mammals to long-established wind farms in India", *Scientific Reports*, Vol. 12/1, https://doi.org/10.1038/s41598-022-05159-1. [176]

Kurek, P., T. Sparks and P. Tryjanowski (2015), "Electricity pylons may be potential foci for the invasion of black cherry Prunus serotina in intensive farmland", *Acta Oecologica*, Vol. 62, pp. 40-44, https://doi.org/10.1016/j.actao.2014.11.005. [285]

Lambert, Q. et al. (2021), "Effects of solar park construction and solar panels on soil quality, microclimate, CO 2 effluxes, and vegetation under a Mediterranean climate", *Land Degradation & Development*, Vol. 32/18, pp. 5190-5202, https://doi.org/10.1002/ldr.4101. [34]

Langhamer, O. (2012), "Artificial Reef Effect in relation to Offshore Renewable Energy Conversion: State of the Art", *The Scientific World Journal*, Vol. 2012, pp. 1-8, https://doi.org/10.1100/2012/386713. [161]

Laurance, W. et al. (2011), "The fate of Amazonian forest fragments: A 32-year investigation", *Biological Conservation*, Vol. 144/1, pp. 56-67, https://doi.org/10.1016/j.biocon.2010.09.021. [292]

Laurance, W., M. Goosem and S. Laurance (2009), "Impacts of roads and linear clearings on tropical forests", *Trends in Ecology & Evolution*, Vol. 24/12, pp. 659-669, https://doi.org/10.1016/j.tree.2009.06.009. [76]

Lèbre, É. et al. (2020), "The social and environmental complexities of extracting energy transition metals", *Nature Communications*, Vol. 11/1, https://doi.org/10.1038/s41467-020-18661-9. [308]

Lee, J. and F. Zhao (2022), *GWEC Global Wind Report 2022*, Global Wind Energy Council, https://gwec.net/wp-content/uploads/2022/04/Annual-Wind-Report-2022_screen_final_April.pdf (accessed on 8 June 2022). [82]

Lehman, R., P. Kennedy and J. Savidge (2007), "The state of the art in raptor electrocution research: A global review", *Biological Conservation*, Vol. 136/2, pp. 159-174, https://doi.org/10.1016/J.BIOCON.2006.09.015. [232]

Leskova, O., R. Frakes and S. Markwith (2022), "Impacting habitat connectivity of the endangered Florida panther for the transition to utility-scale solar energy", *Journal of Applied Ecology*, https://doi.org/10.1111/1365-2664.14098. [49]

Lindeboom, H. et al. (2011), "Short-term ecological effects of an offshore wind farm in the Dutch coastal zone; a compilation", *Environmental Research Letters*, Vol. 6/3, p. 035101, https://doi.org/10.1088/1748-9326/6/3/035101. [158]

Li, S. (ed.) (2017), "High mortality of blue, humpback and fin whales from modeling of vessel collisions on the U.S. West Coast suggests population impacts and insufficient protection", *PLOS ONE*, Vol. 12/8, p. e0183052, https://doi.org/10.1371/journal.pone.0183052. [136]

LOD (2019), *Installation of bird protection measures on the high voltage electricity*, https://www.birds-electrogrid.lt/uploads/Projekto%20dokumentai/LIFE13%20BIO-LT-001303%20FINAL%20REPORT_approved_web.pdf. [245]

Łopucki, R., D. Klich and S. Gielarek (2017), "Do terrestrial animals avoid areas close to turbines in functioning wind farms in agricultural landscapes?", *Environmental Monitoring and Assessment*, Vol. 189/7, https://doi.org/10.1007/S10661-017-6018-Z. [163]

Lóránt, M. and C. Vadász (2014), "The Effect of Above-Ground Medium Voltage Power Lines on Displaying Site Selection of the Great Bustard (<i>Otis Tarda</i>) in Central Hungary", *Ornis Hungarica*, Vol. 22/2, pp. 42-49, https://doi.org/10.2478/orhu-2014-0017. [264]

Loss, S. et al. (2014), "Bird–building collisions in the United States: Estimates of annual mortality and species vulnerability", *The Condor*, Vol. 116/1, pp. 8-23, https://doi.org/10.1650/condor-13-090.1. [103]

Loss, S., T. Will and P. Marra (2015), "Direct Mortality of Birds from Anthropogenic Causes", *Annual Review of Ecology, Evolution, and Systematics*, Vol. 46/1, pp. 99-120, https://doi.org/10.1146/annurev-ecolsys-112414-054133. [222]

Loss, S., T. Will and P. Marra (2013), "Estimates of bird collision mortality at wind facilities in the contiguous United States", *Biological Conservation*, Vol. 168, pp. 201-209, https://doi.org/10.1016/j.biocon.2013.10.007. [94]

Loss, S., T. Will and P. Marra (2012), "Direct human-caused mortality of birds: improving quantification of magnitude and assessment of population impact", *Frontiers in Ecology and the Environment*, Vol. 10/7, pp. 357-364, https://doi.org/10.1890/110251. [226]

Lovich, J. and J. Ennen (2011), "Wildlife Conservation and Solar Energy Development in the Desert Southwest, United States", *BioScience*, Vol. 61/12, pp. 982-992, https://doi.org/10.1525/bio.2011.61.12.8. [40]

Luckeneder, S. et al. (2021), "Surge in global metal mining threatens vulnerable ecosystems", *Global Environmental Change*, Vol. 69, p. 102303, https://doi.org/10.1016/j.gloenvcha.2021.102303. [304]

Lusty, P. et al. (2021), *Deep-sea mining evidence review*, Report commissioned by UK Government., https://www.bgs.ac.uk/download/deep-sea-mining-evidence-review/. [310]

Mahmood, M. et al. (2013), "Effects of electromagnetic field on the nitrogen, protein and chlorophyll
content and peroxidase enzyme activity in oil palm (Elaeis guineensis Jacq.)
leaves", *Emirates Journal of Food and Agriculture*, Vol. 25/6, p. 471, https://doi.org/10.9755/ejfa.v25i6.15583. [279]

Månberger, A. and B. Stenqvist (2018), "Global metal flows in the renewable energy transition: Exploring the effects of substitutes, technological mix and development", *Energy Policy*, Vol. 119, pp. 226-241, https://doi.org/10.1016/j.enpol.2018.04.056. [300]

Margalida, A. (ed.) (2014), "Refining Estimates of Bird Collision and Electrocution Mortality at Power Lines in the United States", *PLoS ONE*, Vol. 9/7, p. e101565, https://doi.org/10.1371/journal.pone.0101565. [221]

Marques, A. et al. (2014), "Understanding bird collisions at wind farms: An updated review on the causes and possible mitigation strategies", *Biological Conservation*, Vol. 179, pp. 40-52, https://doi.org/10.1016/j.biocon.2014.08.017. [119]

Martinez-Alier, J. (2001), "Mining conflicts, environmental justice, and valuation", *Journal of Hazardous Materials*, Vol. 86/1-3, pp. 153-170, https://doi.org/10.1016/s0304-3894(01)00252-7. [321]

Masson, V. et al. (2014), "Solar panels reduce both global warming and urban heat island", *Frontiers in Environmental Science*, Vol. 2, https://doi.org/10.3389/fenvs.2014.00014. [56]

Matthew, F. et al. (2016), *Understanding the risk to European protected species (bats) at onshore wind turbine sites to inform risk management. Technical Report. Defra*, http://eprints.uwe.ac.uk/33789. [130]

Maxwell, S. et al. (2022), "Potential impacts of floating wind turbine technology for marine species and habitats", *Journal of Environmental Management*, Vol. 307, p. 114577, https://doi.org/10.1016/j.jenvman.2022.114577. [86]

May, R. et al. (2020), "Paint it black: Efficacy of increased wind turbine rotor blade visibility to reduce avian fatalities", *Ecology and Evolution*, Vol. 10/16, pp. 8927-8935, https://doi.org/10.1002/ece3.6592. [146]

McClure, C. et al. (2021), "Flight Altitudes of Raptors in Southern Africa Highlight Vulnerability of Threatened Species to Wind Turbines", *Frontiers in Ecology and Evolution*, Vol. 9, https://doi.org/10.3389/fevo.2021.667384. [118]

McGarry, T. (2017), *Understanding the Effectiveness of Acoustic Deterrent Devices on Minke Whale (Balaenoptera acutorostrata), A Low Frequency Cetacean*. [189]

McGarry, T. et al. (2022), *Evidence base for application of Acoustic Deterrent Devices (ADDs) as marine mammal mitigation (Version 3)*, JNCC, https://data.jncc.gov.uk/data/e2d08d7a-998b-4814-a0ae-4edf5d887a02/jncc-report-615-v4.pdf. [188]

McInturff, A. et al. (2020), "Fence Ecology: Frameworks for Understanding the Ecological Effects of Fences", *BioScience*, https://doi.org/10.1093/biosci/biaa103. [51]

Milligan, M. et al. (2023), "Wind-energy development alters pronghorn migration at multiple scales", *Ecology and Evolution*, Vol. 13/1, https://doi.org/10.1002/ece3.9687. [174]

Minderman, J. (ed.) (2015), "Retrofitting of power lines effectively reduces mortality by electrocution in large birds: an example with the endangered Bonelli's eagle", *Journal of Applied Ecology*, Vol. 52/6, pp. 1465-1473, https://doi.org/10.1111/1365-2664.12476. [241]

Montag, H., G. Parker and T. Clarkson (2016), *The Effects of Solar Farms on Local Biodiversity: A Comparative Study*, https://2lwej44565rn2mmjlk31pmwq-wpengine.netdna-ssl.com/wp-content/uploads/2016/05/The-Effects-of-Solar-Farms-on-Local-Biodiversity-A-Comparative-Study.pdf (accessed on 16 December 2021). [36]

Moore, R., K. Nekaris and C. Eschmann (2010), "Habitat use by western purple-faced langurs Trachypithecus vetulus nestor (Colobinae) in a fragmented suburban landscape", *Endangered Species Research*, Vol. 12/3, pp. 227-234, https://doi.org/10.3354/esr00307. [237]

Moore, R., Wihermanto and K. Nekaris (2014), "Compassionate conservation, rehabilitation and translocation of Indonesian slow lorises", *Endangered Species Research*, Vol. 26/2, pp. 93-102, https://doi.org/10.3354/esr00620. [236]

MTE (2022), • *MTE (2022), « L'éolien et les moteurs pour véhicules électriques : choix technologiques, enjeux matières et opportunités industrielles » https://www.ecologie.gouv.fr/sites/default/files/rapport_cgdd_02_eolien_moteurs_vehicules_electriques_juillet2022.p*, https://www.ecologie.gouv.fr/sites/default/files/rapport_cgdd_02_eolien_moteurs_vehicules_electriques_juillet2022.pdf. [301]

MTE (2022), *La mobilité bas-carbone : choix technologiques, enjeux matières et opportunités industrielles*, https://www.ecologie.gouv.fr/sites/default/files/rapport_cgdd_01_mobilite_bas_carbone_fevrier2022.pdf. [297]

MTE (2020), *Le photovoltaïque : choix technologiques, enjeux matières et opportunités industrielles*, http://www.ecologie.gouv.fr/sites/default/files/Plan%20ressources%20Photovoltaique.pdf. [296]

MTE (2020), *Les réseaux électriques (lignes électriques, stockage stationnaire et réseau intelligents) : choix technologiques, enjeux matières et opportunités industrielles*, https://www.ecologie.gouv.fr/sites/default/files/Plan%20ressources%20R%C3%A9. [298]

Murguía, D., S. Bringezu and R. Schaldach (2016), "Global direct pressures on biodiversity by large-scale metal mining: Spatial distribution and implications for conservation", *Journal of Environmental Management*, Vol. 180, pp. 409-420, https://doi.org/10.1016/j.jenvman.2016.05.040. [306]

Murphy-Mariscal, M., S. Grodsky and R. Hernandez (2018), "Solar Energy Development and the Biosphere", *A Comprehensive Guide to Solar Energy Systems*, pp. 391-405, https://doi.org/10.1016/B978-0-12-811479-7.00020-8. [23]

Murphy-Mariscal, M., S. Grodsky and R. Hernandez (2018), "Solar Energy Development and the Biosphere", in *A Comprehensive Guide to Solar Energy Systems*, Elsevier, https://doi.org/10.1016/b978-0-12-811479-7.00020-8. [32]

Murphy-Mariscal, M., S. Grodsky and R. Hernandez (2018), "Solar Energy Development and the Biosphere", *A Comprehensive Guide to Solar Energy Systems*, pp. 391-405, https://doi.org/10.1016/B978-0-12-811479-7.00020-8. [61]

Murray, J. (2021), *Profiling the five largest solar power plants in China*, https://www.nsenergybusiness.com/features/largest-solar-plants-china/ (accessed on 19 July 2022). [8]

N Yenkie, M. et al. (2006), "Impacts of mining activities on water and soil German Academic Exchange Fellowship View project Wastewater treatment View project Impacts of Mining Activities on Water and Soil", *Article in Indian Journal of Environmental Health*, Vol. 47/4, pp. 81-90, https://www.researchgate.net/publication/5932180 (accessed on 1 August 2022). [320]

Nekola, J. (2012), "The impact of a utility corridor on terrestrial gastropod biodiversity", *Biodiversity and Conservation*, Vol. 21/3, pp. 781-795, https://doi.org/10.1007/s10531-011-0216-8. [258]

Nellemann, C. et al. (2001), "Winter distribution of wild reindeer in relation to power lines, roads and resorts", *Biological Conservation*, Vol. 101/3, pp. 351-360, https://doi.org/10.1016/s0006-3207(01)00082-9. [272]

OECD (2023), *Taming Wildfires in the Context of Climate Change*, OECD Publishing, Paris, https://doi.org/10.1787/dd00c367-en. [289]

OECD (2021), "Towards G7 action to combat ghost fishing gear: A background report prepared for the 2021 G7 Presidency of the United Kingdom", *OECD Environment Policy Papers*, No. 25, OECD Publishing, Paris, https://doi.org/10.1787/a4c86e42-en. [138]

OECD (2019), *Global Material Resources Outlook to 2060: Economic Drivers and Environmental Consequences*, OECD Publishing, Paris, https://doi.org/10.1787/9789264307452-en. [295]

OECD (2016), *OECD Due Diligence Guidance for Responsible Supply Chains of Minerals from Conflict-Affected and High-Risk Areas: Third Edition*, OECD Publishing, Paris, https://doi.org/10.1787/9789264252479-en. [299]

O'Shea, T. et al. (2016), "Multiple mortality events in bats: a global review", *Mammal Review*, Vol. 46/3, pp. 175-190, https://doi.org/10.1111/mam.12064. [207]

Parker, L., V. Nijman and K. Nekaris (2008), "When there is no forest left: fragmentation, local extinction, and small population sizes in the Sri Lankan western purple-faced langur", *Endangered Species Research*, Vol. 5, pp. 29-36, https://doi.org/10.3354/esr00107. [238]

Park, J. and Y. Do (2022), "Wind Turbine Noise Behaviorally and Physiologically Changes Male Frogs", *Biology*, Vol. 11/4, p. 516, https://doi.org/10.3390/biology11040516. [187]

Parsons, G. et al. (2020), "The effect of offshore wind power projects on recreational beach use on the east coast of the United States: Evidence from contingent-behavior data", *Energy Policy*, Vol. 144, p. 111659, https://doi.org/10.1016/j.enpol.2020.111659. [197]

Pearce-Higgins, J. et al. (2012), "Greater impacts of wind farms on bird populations during construction than subsequent operation: results of a multi-site and multi-species analysis", *Journal of Applied Ecology*, Vol. 49/2, pp. 386-394, https://doi.org/10.1111/j.1365-2664.2012.02110.x. [172]

Pearce-Higgins, J. et al. (2009), "The distribution of breeding birds around upland wind farms", *Journal of Applied Ecology*, Vol. 46/6, pp. 1323-1331, https://doi.org/10.1111/J.1365-2664.2009.01715.X. [167]

Pearse, A. et al. (2021), "Migrating Whooping Cranes avoid wind-energy infrastructure when selecting stopover habitat", *Ecological Applications*, Vol. 31/5, https://doi.org/10.1002/eap.2324. [168]

Perold, V., S. Ralston-Paton and P. Ryan (2020), "On a collision course? The large diversity of birds killed by wind turbines in South Africa", *Ostrich*, Vol. 91/3, pp. 228-239, https://doi.org/10.2989/00306525.2020.1770889. [100]

Perrow, M. (ed.) (2017), *Vegetation*, Pelagic Publishing. [192]

Peschko, V. et al. (2020), "Effects of offshore windfarms on seabird abundance: Strong effects in spring and in the breeding season", *Marine Environmental Research*, Vol. 162, p. 105157, https://doi.org/10.1016/j.marenvres.2020.105157. [164]

Plewa, R. et al. (2020), "Establishment and Maintenance of Power Lines are Important for Insect Diversity in Central Europe", *Zoological Studies*, Vol. 59, p. e3, https://doi.org/10.6620/ZS.2020.59-3. [259]

Poessel, S. et al. (2018), "Flight response to spatial and temporal correlates informs risk from wind turbines to the California Condor", *The Condor*, Vol. 120/2, pp. 330-342, https://doi.org/10.1650/condor-17-100.1. [116]

Pohlman, C., S. Turton and M. Goosem (2009), "Temporal variation in microclimatic edge effects near powerlines, highways and streams in Australian tropical rainforest", *Agricultural and Forest Meteorology*, Vol. 149/1, pp. 84-95, https://doi.org/10.1016/J.AGRFORMET.2008.07.003. [248]

Pollitt, M. (2009), *Electricity liberalisation in the European Union: A progress report.*. [214]

Pol, M. (ed.) (2022), "Demersal fish and invertebrate catches relative to construction and operation of North America's first offshore wind farm", *ICES Journal of Marine Science*, Vol. 79/4, pp. 1274-1288, https://doi.org/10.1093/icesjms/fsac051. [184]

Popper, A. et al. (2022), "Offshore wind energy development: Research priorities for sound and vibration effects on fishes and aquatic invertebrates", *The Journal of the Acoustical Society of America*, Vol. 151/1, pp. 205-215, https://doi.org/10.1121/10.0009237. [325]

Pruett, C., M. Patten and D. Wolfe (2009), “It’s Not Easy Being Green: Wind Energy and a Declining Grassland Bird”, *BioScience*, Vol. 59/3, pp. 257-262, https://doi.org/10.1525/bio.2009.59.3.10. [265]

Raab, R. et al. (2011), “Underground cabling and marking of power lines: conservation measures rapidly reduced mortality of West-Pannonian Great Bustards <i>Otis tarda</i>”, *Bird Conservation International*, Vol. 22/3, pp. 299-306, https://doi.org/10.1017/s0959270911000463. [246]

Randle-Boggis, R. et al. (2020), “Realising co-benefits for natural capital and ecosystem services from solar parks: A co-developed, evidence-based approach”, *Renewable and Sustainable Energy Reviews*, Vol. 125, p. 109775, https://doi.org/10.1016/j.rser.2020.109775. [74]

Raptor Protection of Slovakia (2019), *Protecting birds from power lines focusing on countries of Danube/Carpathian region*, https://energy.danube-region.eu/wp-content/uploads/sites/6/sites/6/2019/09/Study_Birds_FIN.pdf (accessed on 7 February 2023). [247]

Reznik, G., J. Pires and L. Freitas (2012), “Efeito de bordas lineares na fenologia de espécies arbóreas zoocóricas em um remanescente de Mata Atlântica”, *Acta Botanica Brasilica*, Vol. 26/1, pp. 65-73, https://doi.org/10.1590/s0102-33062012000100008. [250]

Richardson, M. et al. (2017), “A review of the impact of pipelines and power lines on biodiversity and strategies for mitigation”, *Biodiversity and Conservation*, Vol. 26/8, pp. 1801-1815, https://doi.org/10.1007/S10531-017-1341-9/TABLES/3. [287]

Richardson, S. et al. (2021), “Peaks in bat activity at turbines and the implications for mitigating the impact of wind energy developments on bats”, *Scientific Reports*, Vol. 11/1, https://doi.org/10.1038/s41598-021-82014-9. [120]

Rieucau, G., W. Vickery and G. Doucet (2009), “A patch use model to separate effects of foraging costs on giving-up densities: an experiment with white-tailed deer (Odocoileus virginianus)”, *Behavioral Ecology and Sociobiology*, Vol. 63/6, pp. 891-897, https://doi.org/10.1007/s00265-009-0732-7. [271]

Rioux, S., J. Savard and A. Gerick (2013), “Avian mortalities due to transmission line collisions: a review of current estimates and field methods with an emphasis on applications to the Canadian electric network”, *Avian Conservation and Ecology*, Vol. 8/2, https://doi.org/10.5751/ace-00614-080207. [223]

Roehrkasten, S., D. Schaeuble and S. Helgenberger (2015), *Secure and Sustainable Power Generation in a Water-Constrained World*, https://www.rifs-potsdam.de/sites/default/files/files/iass_2015_powergenerationinawaterconstrainedworld.pdf. [43]

Roemer, C. et al. (2017), “Bat flight height monitored from wind masts predicts mortality risk at wind farms”, *Biological Conservation*, Vol. 215, pp. 116-122, https://doi.org/10.1016/j.biocon.2017.09.002. [117]

Rollins, K. et al. (2012), “A Forensic Investigation Into the Etiology of Bat Mortality at a Wind Farm: Barotrauma or Traumatic Injury?”, *Veterinary Pathology*, Vol. 49/2, pp. 362-371, https://doi.org/10.1177/0300985812436745. [91]

RPS (2021), *Electrocutions and Collisions of Birds in EU Countries: The Negative Impact and Best Practices for Mitigation*. [229]

RTE (2019), *French Transmission Network Development Plan*, https://eolmernormandie.debatpublic.fr/images/documents/dmo/biblio/sddr2019_rapport_complet.pdf. [302]

Rydell, J. et al. (2010), "Bat mortality at wind turbines in northwestern Europe", *Acta Chiropterologica*, Vol. 12/2, pp. 261-274, https://doi.org/10.3161/150811010X537846. [113]

Rydell, J. et al. (2010), "Bat Mortality at Wind Turbines in Northwestern Europe", *Acta Chiropterologica*, Vol. 12/2, pp. 261-274, https://doi.org/10.3161/150811010x537846. [124]

Sacramento, E. et al. (2015), "Scenarios for use of floating photovoltaic plants in Brazilian reservoirs", *IET Renewable Power Generation*, Vol. 9/8, pp. 1019-1024, https://doi.org/10.1049/iet-rpg.2015.0120. [13]

Saikia, L. (ed.) (2022), "Land-use intensity of electricity production and tomorrow's energy landscape", *PLOS ONE*, Vol. 17/7, p. e0270155, https://doi.org/10.1371/journal.pone.0270155. [19]

Sánchez-Bayo, F. and K. Wyckhuys (2019), "Worldwide decline of the entomofauna: A review of its drivers", *Biological Conservation*, Vol. 232, pp. 8-27, https://doi.org/10.1016/J.BIOCON.2019.01.020. [134]

Santiago-Quesada, F. et al. (2014), "Roost location and landscape attributes influencing habitat selection of migratory waterbirds in rice fields", *Agriculture, Ecosystems & Environment*, Vol. 188, pp. 97-102, https://doi.org/10.1016/j.agee.2014.02.019. [268]

Schaub, M. et al. (2010), "Massive immigration balances high anthropogenic mortality in a stable eagle owl population: Lessons for conservation", *Biological Conservation*, Vol. 143/8, pp. 1911-1918, https://doi.org/10.1016/j.biocon.2010.04.047. [225]

Schaub, M. and R. Pradel (2004), "ASSESSING THE RELATIVE IMPORTANCE OF DIFFERENT SOURCES OF MORTALITY FROM RECOVERIES OF MARKED ANIMALS", *Ecology*, Vol. 85/4, pp. 930-938, https://doi.org/10.1890/03-0012. [224]

Schmuecker, S. et al. (2020), "Use of Upland and Riparian Areas by Wintering Bald Eagles and Implications for Wind Energy", *The Journal of Wildlife Management*, Vol. 84/8, pp. 1578-1589, https://doi.org/10.1002/jwmg.21927. [128]

Schöll, E. and U. Nopp-Mayr (2021), "Impact of wind power plants on mammalian and avian wildlife species in shrub- and woodlands", *Biological Conservation*, Vol. 256, p. 109037, https://doi.org/10.1016/j.biocon.2021.109037. [326]

Scholz, C. and C. Voigt (2022), "Diet analysis of bats killed at wind turbines suggests large-scale losses of trophic interactions", *Conservation Science and Practice*, p. e12744, https://doi.org/10.1111/CSP2.12744. [203]

Scottish Government (2020), *Renewable developments on Scotland's national forests and lands: EIR release*, https://www.gov.scot/publications/foi-19-02646/. [153]

SEER (2022), *Benthic Disturbance from offshore Wind Foundations, Anchors, and Cables*, Tethys, https://tethys.pnnl.gov/summaries/benthic-disturbance-offshore-wind-foundations-anchors-cables. [87]

Shaffer, J. and D. Buhl (2015), "Effects of wind-energy facilities on breeding grassland bird distributions", *Conservation Biology*, Vol. 30/1, pp. 59-71, https://doi.org/10.1111/cobi.12569. [166]

Shobrak, M. (2012), "Electrocution and collision of birds with power lines in Saudi Arabia", *Zoology in the Middle East*, Vol. 57/1, pp. 45-52, https://doi.org/10.1080/09397140.2012.10648962. [227]

Silva, J. et al. (2010), "Estimating the influence of overhead transmission power lines and landscape context on the density of little bustard Tetrax tetrax breeding populations", *Ecological Modelling*, Vol. 221/16, pp. 1954-1963, https://doi.org/10.1016/j.ecolmodel.2010.03.027. [269]

Skarin, A., P. Sandström and M. Alam (2018), "Out of sight of wind turbines—Reindeer response to wind farms in operation", *Ecology and Evolution*, Vol. 8/19, pp. 9906-9919, https://doi.org/10.1002/ECE3.4476. [175]

Skov, H. et al. (2018), *Bird Collision Avoidance Study*, http://www.nirasconsulting.co.uk (accessed on 1 July 2022). [112]

Smallwood, K. (2013), "Comparing bird and bat fatality-rate estimates among North American wind-energy projects", *Wind Energy and Wildlife Conservation*, Vol. 37/1, pp. 19-33, https://doi.org/10.1002/wsb.260. [96]

Smallwood, K. (2013), "Comparing bird and bat fatality-rate estimates among North American wind-energy projects", *Wildlife Society Bulletin*, Vol. 37/1, pp. 19-33, https://doi.org/10.1002/wsb.260. [105]

Smallwood, K. and D. Bell (2020), "Effects of Wind Turbine Curtailment on Bird and Bat Fatalities", *The Journal of Wildlife Management*, Vol. 84/4, pp. 685-696, https://doi.org/10.1002/jwmg.21844. [148]

Smallwood, K. and B. Karas (2009), "Avian and Bat Fatality Rates at Old-Generation and Repowered Wind Turbines in California", *Journal of Wildlife Management*, Vol. 73/7, pp. 1062-1071, https://doi.org/10.2193/2008-464. [131]

Smallwood, K. and C. Thelander (2008), "Bird Mortality in the Altamont Pass Wind Resource Area, California", *Journal of Wildlife Management*, Vol. 72/1, pp. 215-223, https://doi.org/10.2193/2007-032. [147]

Smith, K. et al. (2020), "Pronghorn Winter Resource Selection Before and After Wind Energy Development in South-Central Wyoming", *Rangeland Ecology & Management*, Vol. 73/2, pp. 227-233, https://doi.org/10.1016/j.rama.2019.12.004. [173]

Smith, M. et al. (2008), "Mammalian Predator Distribution Around a Transmission Line", *Southeastern Naturalist*, https://bioone.org/journals/Southeastern-Naturalist/volume-7/issue-2/1528-7092(2008)7[289:MPDAAT]2.0.CO;2/Mammalian-Predator-Distribution-Around-a-Transmission-Line/10.1656/1528-7092(2008)7[289:MPDAAT]2.0.CO;2.short (accessed on 2 August 2022). [262]

SolarPACES (2018), *How CSP Works: Tower, Trough, Fresnel or Dish - SolarPACES*, https://www.solarpaces.org/how-csp-works/ (accessed on 16 December 2021). [15]

Sonter, L., S. Ali and J. Watson (2018), “Mining and biodiversity: key issues and research needs in conservation science”, *Proceedings of the Royal Society B: Biological Sciences*, Vol. 285/1892, p. 20181926, https://doi.org/10.1098/rspb.2018.1926. [313]

Sonter, L. et al. (2020), “Renewable energy production will exacerbate mining threats to biodiversity”, *Nature Communications 2020 11:1*, Vol. 11/1, pp. 1-6, https://doi.org/10.1038/s41467-020-17928-5. [307]

Sonter, L. et al. (2017), “Mining drives extensive deforestation in the Brazilian Amazon”, *Nature Communications*, Vol. 8/1, https://doi.org/10.1038/s41467-017-00557-w. [323]

Sonter, L. et al. (2014), “Processes of land use change in mining regions”, *Journal of Cleaner Production*, Vol. 84, pp. 494-501, https://doi.org/10.1016/j.jclepro.2014.03.084. [315]

Sovacool, B. (2009), “Contextualizing avian mortality: A preliminary appraisal of bird and bat fatalities from wind, fossil-fuel, and nuclear electricity”, *Energy Policy*, Vol. 37/6, pp. 2241-2248, https://doi.org/10.1016/j.enpol.2009.02.011. [102]

Suuronen, A. et al. (2017), “The Influence of Solar Power Plants on Microclimatic Conditions and the Biotic Community in Chilean Desert Environments”, *Environmental Management*, Vol. 60/4, pp. 630-642, https://doi.org/10.1007/s00267-017-0906-4. [59]

Tabassum-Abbasi, M. et al. (2014), “Wind energy: Increasing deployment, rising environmental concerns”, *Renewable and Sustainable Energy Reviews*, Vol. 31, pp. 270-288, https://doi.org/10.1016/j.rser.2013.11.019. [152]

Taha, H. (2013), “The potential for air-temperature impact from large-scale deployment of solar photovoltaic arrays in urban areas”, *Solar Energy*, Vol. 91, pp. 358-367, https://doi.org/10.1016/j.solener.2012.09.014. [55]

Tanner, A. and A. Johnston (2017), “The Impact of Rural Electric Access on Deforestation Rates”, *World Development*, Vol. 94, pp. 174-185, https://doi.org/10.1016/j.worlddev.2016.12.046. [2]

Tanner, K., K. Moore and B. Pavlik (2014), “Changes to shade and water regimes imposed by solar development affect desert plant performance and community attributes”, *Conference: 99th ESA Annual Convention*, https://www.researchgate.net/publication/267287102_Changes_to_shade_and_water_regimes_imposed_by_solar_development_affect_desert_plant_performance_and_community_attributes. [60]

ter Hofstede, R. et al. (2022), “Offshore wind farms contribute to epibenthic biodiversity in the North Sea”, *Journal of Sea Research*, Vol. 185, p. 102229, https://doi.org/10.1016/j.seares.2022.102229. [157]

Thaker, M., A. Zambre and H. Bhosale (2018), “Wind farms have cascading impacts on ecosystems across trophic levels”, *Nature Ecology & Evolution*, Vol. 2/12, pp. 1854-1858, https://doi.org/10.1038/s41559-018-0707-z. [202]

Thaxter, C. et al. (2017), “Bird and bat species’ global vulnerability to collision mortality at wind farms revealed through a trait-based assessment”, *Proceedings of the Royal Society*, Vol. 284/1862, https://doi.org/10.1098/rspb.2017.0829. [108]

Therkildsen, O. et al. (2021), “Changes in flight paths of large-bodied birds after construction of large terrestrial wind turbines”, *Journal of Environmental Management*, Vol. 290, p. 112647, https://doi.org/10.1016/j.jenvman.2021.112647. [169]

Thompson, D. (ed.) (2016), “Response of migrating raptors to an increasing number of wind farms”, *Journal of Applied Ecology*, Vol. 53/6, pp. 1667-1675, https://doi.org/10.1111/1365-2664.12673. [156]

Thompson, M. et al. (2017), “Factors associated with bat mortality at wind energy facilities in the United States”, *Biological Conservation*, Vol. 215, pp. 241-245, https://doi.org/10.1016/J.BIOCON.2017.09.014. [123]

Tintó, A., J. Real and S. Mañosa (2010), “Predicting and Correcting Electrocution of Birds in Mediterranean Areas”, *Journal of Wildlife Management*, Vol. 74/8, pp. 1852-1862, https://doi.org/10.2193/2009-521. [231]

Tomás, G. et al. (2012), “Clutch size and egg volume in great tits (Parus major) increase under low intensity electromagnetic fields: A long-term field study”, *Environmental Research*, Vol. 118, pp. 40-46, https://doi.org/10.1016/j.envres.2012.07.007. [276]

Tomé, R. et al. (2017), “Radar Assisted Shutdown on Demand Ensures Zero Soaring Bird Mortality at a Wind Farm Located in a Migratory Flyway”, in *Wind Energy and Wildlife Interactions*, Springer International Publishing, Cham, https://doi.org/10.1007/978-3-319-51272-3_7. [149]

Turney, D. and V. Fthenakis (2011), “Environmental impacts from the installation and operation of large-scale solar power plants”, *Renewable and Sustainable Energy Reviews*, Vol. 15/6, pp. 3261-3270, https://doi.org/10.1016/j.rser.2011.04.023. [31]

Urziceanu, M. et al. (2021), “Local-scale impact of wind energy farms on rare, endemic, and threatened plant species”, *PeerJ*, Vol. 9, p. e11390, https://doi.org/10.7717/peerj.11390. [154]

van de Ven, D. et al. (2021), “The potential land requirements and related land use change emissions of solar energy”, *Scientific Reports*, Vol. 11/1, https://doi.org/10.1038/s41598-021-82042-5. [69]

van der Weijde, A. and B. Hobbs (2012), “The economics of planning electricity transmission to accommodate renewables: Using two-stage optimisation to evaluate flexibility and the cost of disregarding uncertainty”, *Energy Economics*, Vol. 34/6, pp. 2089-2101, https://doi.org/10.1016/j.eneco.2012.02.015. [213]

Van Hoey, G. et al. (2021), *Overview of the effects of offshore wind farms on fisheries and aquaculture*, Publications Office of the European Union. [199]

Velilla, E. et al. (2021), “Vibrational noise from wind energy-turbines negatively impacts earthworm abundance”, *Oikos*, Vol. 130/6, pp. 844-849, https://doi.org/10.1111/oik.08166. [177]

Visser, E. et al. (2019), “Assessing the impacts of a utility-scale photovoltaic solar energy facility on birds in the Northern Cape, South Africa”, *Renewable Energy*, Vol. 133, pp. 1285-1294, https://doi.org/10.1016/j.renene.2018.08.106. [26]

Voigt, C. (2021), “Insect fatalities at wind turbines as biodiversity sinks”, *Conservation Science and Practice*, Vol. 3/5, p. e366, https://doi.org/10.1111/CSP2.366. [132]

Voigt, C. et al. (2022), "Wind turbines without curtailment produce large numbers of bat fatalities throughout their lifetime: A call against ignorance and neglect", *Global Ecology and Conservation*, Vol. 37, p. e02149, https://doi.org/10.1016/j.gecco.2022.e02149. [107]

Voigt, C. et al. (2015), "Wildlife and renewable energy: German politics cross migratory bats", *European Journal of Wildlife Research*, Vol. 61/2, pp. 213-219, https://doi.org/10.1007/s10344-015-0903-y. [106]

Voltaire, L. and O. Koutchade (2020), "Public acceptance of and heterogeneity in behavioral beach trip responses to offshore wind farm development in Catalonia (Spain)", *Resource and Energy Economics*, Vol. 60, p. 101152, https://doi.org/10.1016/j.reseneeco.2020.101152. [196]

Votier, S. (ed.) (2016), "Avoidance of wind farms by harbour seals is limited to pile driving activities", *Journal of Applied Ecology*, Vol. 53/6, pp. 1642-1652, https://doi.org/10.1111/1365-2664.12678. [181]

Walston, L. et al. (2021), "Modeling the ecosystem services of native vegetation management practices at solar energy facilities in the Midwestern United States", *Ecosystem Services*, Vol. 47, p. 101227, https://doi.org/10.1016/j.ecoser.2020.101227. [75]

Walston, L. et al. (2016), "A preliminary assessment of avian mortality at utility-scale solar energy facilities in the United States", *Renewable Energy*, Vol. 92, pp. 405-414, https://doi.org/10.1016/j.renene.2016.02.041. [21]

Walston, L. et al. (2016), "A preliminary assessment of avian mortality at utility-scale solar energy facilities in the United States", *Renewable Energy*, Vol. 92, pp. 405-414, https://doi.org/10.1016/J.RENENE.2016.02.041. [27]

Whitehead, A., H. Kujala and B. Wintle (2017), *Dealing with Cumulative Biodiversity Impacts in Strategic Environmental Assessment: A New Frontier for Conservation Planning*, Wiley-Blackwell, https://doi.org/10.1111/conl.12260. [3]

Wilson (2006), *Collision risks between marine renewable energy devices and mammals, fish and diving birds: Report to the Scottish executive*. [137]

Wilson, M. et al. (2015), "Habitat fragmentation and biodiversity conservation: key findings and future challenges", *Landscape Ecology*, Vol. 31/2, pp. 219-227, https://doi.org/10.1007/s10980-015-0312-3. [44]

Wilson, R., H. Marsh and J. Winter (2007), "Importance of canopy connectivity for home range and movements of the rainforest arboreal ringtail possum (Hemibelideus lemuroides)", *Wildlife Research*, Vol. 34/3, p. 177, https://doi.org/10.1071/wr06114. [252]

Wiser, R. et al. (2021), "Land-Based Wind Market Report: 2021 Edition", http://www.osti.gov (accessed on 27 October 2021). [324]

Wu, Z. et al. (2014), "Environmental impacts of large-scale CSP plants in northwestern China", *Environ. Sci.: Processes Impacts*, Vol. 16/10, pp. 2432-2441, https://doi.org/10.1039/c4em00235k. [57]

WWF (2021), *In too deep: What we know, and don't know, about deep seabed mining*, https://wwfint.awsassets.panda.org/downloads/in_too_deep_what_we_know_and_don_t_know_about_deep_seabed_mining_1.pdf. [311]

Yang, X. et al. (2013), "China's ion-adsorption rare earth resources, mining consequences and preservation", *Environmental Development*, Vol. 8, pp. 131-136, https://doi.org/10.1016/j.envdev.2013.03.006. [314]

Zhang, J. (ed.) (2020), "A summary of bird mortality at photovoltaic utility scale solar facilities in the Southwestern U.S.", *PLOS ONE*, Vol. 15/4, p. e0232034, https://doi.org/10.1371/journal.pone.0232034. [24]

Zimmerling, J. and C. Francis (2016), "Bat mortality due to wind turbines in Canada", *The Journal of Wildlife Management*, Vol. 80/8, pp. 1360-1369, https://doi.org/10.1002/jwmg.21128. [206]

Zimmerling, J. et al. (2013), "Canadian Estimate of Bird Mortality Due to Collisions and Direct Habitat Loss Associated with Wind Turbine Developments", *Avian Conservation and Ecology*, Vol. 8/2, https://doi.org/10.5751/ace-00609-080210. [99]

Notes

[1] For sites with multiple full study years, each year was treated as a separate study and is indicated by year in the analyses. Each study is referred to as a site-year.

[2] The cause of mortality could not be determined for 61% of carcasses.

[3] A pre-construction, plant conservation decision that designated buffer zones around rare desert plants within the solar field, which were roped off and left undisturbed.

[4] Mortality estimates of individual studies vary in how raw counts are adjusted for known sources of detection error and sampling intensity, for example, carcass removal by predators or the distance from the wind turbine that is searched.

[5] See (Popper et al., 2022[325]) for a list of short-term priorities for fish and marine invertebrates identified by a workgroup of the 2020 State of the Science Workshop on Wildlife and Offshore Wind Energy.

[6] (Schöll and Nopp-Mayr, 2021[326]) for example, note a relative absence of studies looking at shrub and woodland species, despite woodland being increasingly targeted for wind energy development.

[7] Aluminium and steel are not included in this estimate.

[8] Tropical and subtropical moist broadleaf forest; tropical and subtropical grasslands, savannahs and shrublands; deserts and xeric shrublands; tropical and subtropical dry broadleaf forests; and montane grasslands and shrublands.

[9] Species-richness refers to the number of species in an area.

4 Mainstreaming biodiversity into low-emissions pathways and power sector planning

This chapter examines opportunities, good practices and tools for mainstreaming biodiversity into low-emissions pathways and power system planning. It discusses the importance of low-energy demand pathways, accounting for biodiversity in decisions on electricity generation portfolios and strategically siting renewable power infrastructure. The chapter also discusses the role of institutional and cross-border co-ordination in safeguarding biodiversity while scaling up renewable power.

Since the adoption of the Paris Agreement, national governments have set emission reduction targets in their Nationally Determined Contributions (NDCs). Increasingly, countries are also developing long-term low greenhouse gas emission development strategies (LT-LEDS). Decarbonising energy use through electrification and the expansion of low-emissions electricity generation (particularly from renewable sources) is an important part of governments' NDCs and LT-LEDS. In parallel, governments are pursuing the objective of affordable and reliable energy provision for all, which is the focus of Sustainable Development Goal 7.

Planning for the transition to low-emissions electricity involves setting targets, developing strategies and adopting policies that guide investment in the energy sector. It is often informed by analysis of modelling-based energy scenarios, which illustrate potential pathways for meeting future energy demand and achieving emission reduction targets. Long-term planning is necessary to ensure investments are consistent with the Paris Agreement temperature goals and other long-term policy objectives. As infrastructure is long-lived, decisions made today on energy infrastructure investment have implications for future decades. Long-term plans, such as LT-LEDS, help to orientate mid-term planning (e.g., multi-annual energy programmes, ten-year electricity network plans; ten-year integrated national energy and climate plans; spatial plans), and short-term planning (i.e., decisions dealing with immediate priorities and actions, such as project siting).

As governments plan the low emissions transition, they have an opportunity to design power systems that deliver better outcomes for both climate and biodiversity. Capitalising on this opportunity requires a systemic and integrated approach to electricity planning in which climate, energy and biodiversity (and other societal goals) are considered together from the earliest stages of planning. Such an approach is vital if governments are to achieve their national biodiversity objectives and international commitments, for example, under the Convention on Biological Diversity's Kunming-Montreal Global Biodiversity Framework and the Convention on Migratory Species, as well as the goals of the Paris Agreement and the 2030 Agenda for Sustainable Development (Annex A).

Three key questions confront governments as they plan the transition to low-emissions electricity systems: How much electricity generation capacity is required over time (i.e., electricity demand)? What is the appropriate technology mix to meet this demand? Where should this infrastructure be sited? The response to each of these questions has implications for biodiversity. Yet, the extent to which biodiversity is considered when addressing these questions remains limited. Failure to account for biodiversity in the design and appraisal of energy policies, plans and programmes could lead to governments pursuing unsustainable pathways.

This chapter examines each of these questions in turn, discussing their implications for biodiversity. While the three questions are discussed individually, they are interdependent. For example, changes to energy demand or siting constraints can alter the optimum balance of different renewable power technologies (Wu et al., 2019[1]). The chapter then presents decision support tools that planners and policy makers can use to strengthen biodiversity considerations in decision making. Finally, it discusses the role of institutional co-ordination and cross-border collaboration in delivering a biodiversity-aligned transition to low-emissions electricity systems.

4.1. Biodiversity and energy demand

Electrification coupled with expansion in renewable power is necessary to transition away from fossil fuel-based economies (see Chapter 2). Electricity demand will and must increase. However, the extent of electricity demand growth will depend in part on overall energy demand, which in turn, depends on societal choices and how end-use sectors are designed. How much energy is required will influence the spatial footprint of power systems in the coming decades, thereby determining the extent of competition for land and the risks to biodiversity. A study of France's potential low-carbon energy pathways for 2050, for example, found that mineral resource requirements would be reduced by approximately 20% in a sufficiency (low-demand) scenario[1] compared to a reference scenario,[2] including a 30 million tonne decline for the electrical system and 10 million tonne decline in material for batteries, which could reduce mining pressure on biodiversity (RTE, 2022[2]).

The value of adopting low-emission development pathways based on low-energy (and material) demand is clear. Low-energy demand scenarios can benefit multiple SDGs, including those on biodiversity (SDGs 14 and 15) (Grubler et al., 2018[3]), while pathways that do not have an absolute reduction in energy consumption are likely to make sustainable land management impossible (Tran and Egermann, 2022[4]). The sixth assessment report (AR6) of the Intergovernmental Panel on Climate Change (IPCC, 2022[5]) examines mitigation pathways and land-use implications (Box 4.1). It states that "many challenges, such as dependence on CDR [carbon dioxide removal], pressure on land and biodiversity (e.g., bioenergy) [...] are significantly reduced in modelled pathways that assume using resources more efficiently (e.g., IMP-LD) or that shift global development towards sustainability (e.g., IMP-SP) (high confidence)". Further, the report underscores that "[d]ecent living standards, which encompasses many SDG dimensions, are achievable at lower energy use than previously thought (high confidence)."

Box 4.1. IPCC's sixth assessment report (AR6), mitigation pathways and land-use implications

The AR6 presents five Illustrative Mitigation Pathways (IPCC, 2022[6]). Three of these have >50% likelihood of limiting warming to 1.5°C with no or limited overshoot: IMP-REN, which places greater emphasis on renewables, IMP-LD, which emphasises efficient resource use and shifts in consumption patterns leading to low resource demand, and IMP-SP, which emphasises shifting global pathways to sustainable development, including by reducing inequality. All three of these pathways are characterised by steep early reductions in emissions and a relatively small contribution of net negative emissions, thereby reducing reliance on bioenergy with carbon capture and storage (BECCs), which is land-use intensive and associated with high risks to biodiversity. An emphasis on demand reduction, as illustrated in IM-LD could reduce the need for renewable power infrastructure compared to IMP-REN, while still achieving the 1.5°C goal.

Source: (IPCC, 2022[5]), Climate Change 2022: Mitigation of Climate Change. Working Group III contribution to the Sixth Assessment Report of the Intergovernmental Panel on Climate Change; (Hof et al., 2018[6]), Bioenergy cropland expansion may offset positive effects of climate change mitigation for global vertebrate diversity, 10.1073/pnas.1807745115; (Smith et al., 2019[7]), Land-Management Options for Greenhouse Gas Removal and Their Impacts on Ecosystem Services and the Sustainable Development Goals 10.1146/annurev-environ-101718-033129; (Creutzig et al., 2021[8]), Considering sustainability thresholds for BECCS in IPCC and biodiversity assessments, 10.1111/gcbb.12798.

Pursuing pathways with low energy and material demand will require countries to fully leverage a range of demand-side mitigation measures to improve energy efficiency and change consumer behaviour (IEA, 2021[9]; Creutzig et al., 2021[10]). Technological strategies and innovations are critical but may be insufficient; accompanying social innovations and strategies that address cultures, institutions and practices of energy use and supply are also needed (Tran and Egermann, 2022[4]). Adopting a systems approach (Box 4.2) to development planning could help countries reduce energy demand in end-use systems (e.g. transport and agriculture), thereby delivering climate and other well-being outcomes such as biodiversity (OECD, 2019[11]; OECD, 2021[12]).

Box 4.2. Systems innovation for net zero

Systems innovation for net-zero is a process which allows governments to think innovatively and design climate strategies with the potential to accelerate climate change mitigation while improving wider well-being outcomes. Building on systems thinking, the process unleashes emission reduction opportunities through policies targeted at redesigning systems, often absent or at the margin of current climate strategies. By defining outcomes in terms of well-being (e.g., biodiversity, health, affordability, equity) and making these outcomes central criteria guiding systems' redesign, it mainstreams well-being considerations into the decision-making process of climate strategies from the onset.

Climate action could be more efficient and effective if focused on systems as a whole, so that – by design – end-use systems require less energy and materials, and emit less greenhouse gases, while achieving wider well-being outcomes, such as improving our health and safety, and subsequently better lives. OECD work on the transport sector, for example, identified three vicious cycles in the transport system design that fostered growing car use, and thus led to an energy-intensive transport system. These cycles (induced car demand, urban sprawl and the sustainable modes low-quality trap) are key drivers of car dependency, high energy demand and emissions, as well as other undesirable results (e.g., unequal access to opportunities, poor air quality and limited physical activity). Examples of policies and actions that can reverse these dynamics and reduce the demand of energy-intensive transport modes include wide-scale reallocation and redesign of streets, such as in the case of superblocks; rethinking urban regeneration and new development through the 15' city framework; and mainstreaming on-demand shared services to complement public transport.

Current climate strategies for the transport sector place today an overriding focus on vehicle electrification, leaving car dependency and high energy demand unaddressed. From a well-being lens, electrification of vehicles will be key, but its potential to achieve net-zero goals and contribute to wider well-being objectives will only be fulfilled if vehicle technology improvements are embedded in a wider process of systemic redesign that addresses issues such as car-dependency.

Source: (OECD, 2021[12]), *Transport Strategies for Net-Zero Systems by Design*, OECD Publishing, Paris, https://doi.org/10.1787/0a20f779-en.

4.2. Biodiversity and electricity generation portfolios

Different combinations of renewable power and flexibility technologies (e.g., batteries) may be consistent with the low-emissions electricity transition. However, the choice of technologies will partly determine the biodiversity footprint of transitioning to low-emissions electricity (Luderer et al., 2019[13]). Land and sea-use demands from renewables vary due to the distribution of energy resources, differing power densities of energy technologies and grid infrastructure requirements, among other things. For example, bioenergy is the most land-use intensive of all electricity sources requiring more space per megawatt/hour (MWh) than solar or wind (Lovering et al., 2022[14]). Wind and solar power are more land-use efficient but scaling up these technologies may demand more electricity transmission and energy storage infrastructure. Additionally, some technologies more profoundly impact the species and ecosystems where they are deployed. For example, traditional hydropower schemes tend to be relatively destructive to both terrestrial and freshwater ecosystems (Nilsson et al., 2005[15]; Pörtner et al., 2021[16]). The impact of a given technology mix will differ across countries, according to various factors such as the overlap between energy resources and ecologically important or sensitive areas.

A key challenge for electricity planners and regulators is to identify and deliver the optimal capacity expansion of different electricity generation technologies. Decisions on which electricity generation technologies to invest in, and when, are informed by energy models (e.g., capacity expansion models). These models are developed based on assumptions of demand, technology costs and performance, resource availability and climate regulations (van Ouwerkerk et al., 2022[17]; Wu et al., 2019[1]). The risks to nature and associated costs to society (other than through a climate lens) tend not to be considered.

A major limitation of most capacity expansion modelling is that they are spatially coarse (or aspatial). They therefore do not provide the necessary level of detail to account for potential conflicts between renewable power projects and biodiversity or other environmental, cultural and social constraints. This can create a disconnect between long-term planning and the deployment of renewable power projects (Wu et al., 2019[1]). Long-term planning that does not account for biodiversity impacts and spatial constraints could increase the risk of land-use conflicts, undermining renewable power projects and threatening biodiversity.

To better account for biodiversity in long-term planning, power sector modally could integrate spatially explicit data on biodiversity. For example, scenarios for electricity expansion could be developed that vary constraints on land/sea availability for electricity infrastructure based on the location of areas important for biodiversity protection. This would enable planners to identify power or broader energy sector scenarios that best minimise costs while reducing emissions and protecting biodiversity (Box 4.3). While progress has been made in power sector modelling to better reflect siting constraints and provide more spatially specific outputs (Mai et al., 2021[18]), most studies are limited in scope (geography and/or technologies considered) or do not consider habitat loss.

Another key entry point for biodiversity is the appraisal of different technology options or scenarios for capacity expansion. Decision makers could apply strategic environmental assessments (see SEA section, Chapter 5) to evaluate the cumulative environmental effects of different scenarios. They could also integrate biodiversity criteria and proxies such as land-use intensity of energy (Lovering et al., 2022[14]) into cost-benefit analysis and multi-criteria analyses to identify optimal portfolios (see 4.2).

What constitutes an optimal portfolio depends on a variety of environmental, geographic, social and economic factors. Some jurisdictions may have greater opportunities for siting low-risk renewable power infrastructure. For example, in a scenario with high rooftop solar (an additional 9GW compared to the baseline 2050 forecast), utility-scale capacity build-out to meet California's energy demand could be reduced by 3-6%, which is the equivalent of 200 – 445 km^2 (see 4.4). However, the potential role for rooftop solar varies depending on rooftop availability (e.g., heritage regulations have prevented rooftop solar expansion in some areas), geographical or climatic factors and overall consumption. Estimates indicate that as little as 0.3-1% of Finland's total electricity demand could be met by rooftop PV compared to 5.2-17.4% in Australia and 27.8-92.7% in Indonesia (Capellán-Pérez, de Castro and Arto, 2017[19]).

The potential for conflict across renewable energy and biodiversity targets will be more pronounced in some countries or regions, due to their greater biodiversity and the overlap of areas of high energy resource potential with biodiversity. For example, Central America has a relatively higher overlap of renewable power resources with biodiversity compared to other regions: an estimated 77% of wind energy potential and 75% of solar energy potential is concentrated within the most biodiversity areas, many of which are unprotected (Santangeli et al., 2015[20]). While conflicts can be reduced by integrating biodiversity into capacity expansion scenarios, significant trade-offs may persist. These trade-offs could be partly addressed in later stages of planning and project development, for example by avoiding impacts through project (micro-)siting decisions, minimising impacts through physical controls, restoring project sites and offsetting residual impacts.

Box 4.3. Assessing the biodiversity impact of energy portfolios

Power of Place: Land Conservation and Clean Energy Pathways for California

The Power of Place study developed a planning framework integrating spatially-explicit biodiversity data with an electricity sector capacity expansion model to assess how siting constraints to avoid impacts on natural and working lands in the Western United States could affect technology choices, generation and transmission capacity, system costs, and environmental impacts of pathways that achieve climate targets. The study developed sixty-one scenarios consistent with an 80% reduction in greenhouse gas (GHG) emissions in 2050 compared to 1990. From the scenarios it was concluded that the balance between wind, solar PV, and storage capacity – and the resultant costs – are sensitive to land protections and whether California has access to renewable power from other western states. Applying land protections effectively avoid environmental impacts while achieving GHG targets but can increase system costs (this does not account for avoided costs of cancelled/delayed projects resulting from conflict with biodiversity in scenarios without land protections). The higher costs can be more than offset by allowing access to out-of-state wind and solar resources. In other words, California can reduce costs and improve conservation outcomes by pursuing regional renewable resource development and trade. However, this approach requires significantly more transmission infrastructure and can have greater land use impacts under scenarios with lower levels of environmental protection.

Low to medium ecological risk energy portfolios in the UK

An analysis of potential UK energy portfolios using the DECC 2050 Calculator[3] demonstrated how climate targets could be met with low to medium ecological risk, by excluding high-ecological-risk technologies and carefully siting medium-ecological-risk technologies. The study excluded energy technologies judged to have high risk (e.g., new large-scale hydropower schemes and certain tidal range technologies) and focused instead on what the authors considered to be medium-risk technologies (onshore wind, bioenergy, solar PV, offshore wind, wave power and tidal stream). The analysis evaluated the technical opportunity for each medium-risk technologies, and the physical, policy and ecological constraints limiting their deployment. Technology-specific ecological sensitivity maps were developed based on available information and inferences about the impacts of each technology. This allowed a comparison of the overlap of potential areas for the deployment of specific renewable energy technologies with species or habitats that they may negatively impact. The study identified a low ecological risk scenario for the medium-risk technologies that excludes areas of both high and medium ecological sensitivity from the areas remaining after the various constraints had been applied. The study also identified a medium ecological risk scenario that excludes only areas of high ecological sensitivity from the remaining eligible areas.

Source: (Gove et al., 2016[21]), Reconciling Biodiversity Conservation and Widespread Deployment of Renewable Energy Technologies in the UK, 10.1371/journal.pone.0150956; (Wu et al., 2019[1]), Power of Place Land Conservation and Clean Energy Pathways for California,

As renewable power generation and storage technologies with relatively lower ecological risk become increasingly cost-competitive, countries can reduce their dependence on more harmful forms of electricity generation – fossil fuels but also more harmful forms of low-emission electricity generation. For example, investment in wind, solar and batteries is reducing the need for new hydropower schemes, thereby avoiding negative impacts on rivers[4] and the fragmentation of tens to hundreds of thousands of kilometres of free-flowing rivers globally (Opperman et al., 2019[22]). Hydropower will continue to play an important role in low-emissions electricity pathways for some countries. However, due to wind and solar energy expansion, the focus can shift from traditional dams with significant biodiversity impacts to lower impact schemes that provide storage capability and flexibility to facilitate the integration of variable renewable energies (e.g. strategically-sited off-channel pumped storage) (Opperman et al., 2019[22]; Moran et al., 2018[23]).

Supporting new electricity generation and storage technologies through to full commercialisation may facilitate low-ecological-impact expansion of the electricity system. Increasing the diversity of technologies available for generating low-emissions electricity could provide greater siting flexibility, allowing for areas of low ecological sensitivity to be exploited, while also helping to address concerns with energy variability (McManamay, Vernon and Jager, 2021[24]). For example, the recent development of large offshore floating wind turbines may reduce pressure on terrestrial and coastal ecosystems and provide an opportunity to harness substantial energy resources in areas of low ecological risk. Wave power and tidal stream technologies have not yet been fully commercialised but could also offer an opportunity for low-ecological risk electricity generation (although their ecological impacts will require careful assessment and monitoring, and siting will remain critical).

Additionally, continued investment to refine existing technologies is important for reducing the impact of the power sector on biodiversity. For example, technological developments that improve the power density or efficiency of commercialised technologies such as solar and wind may help reduce their overall spatial requirements. Changes to their design could reduce impacts associated with their construction and operation (e.g., bird-safe power lines designs to reduce electrocution risk, see Chapter 3), or reduce upstream impacts associated with mining for construction materials by improving material efficiency or finding alternative materials (see also Box 5.1, Chapter 5).

Technology development makes it possible to repower old existing solar and wind power facilities to increase their efficiency and capacity. Repowering can increase electricity generation without the downstream[5] land-use/sea-use change associated with new projects. Additionally, it may provide an opportunity to redesign facilities so that they are more biodiversity-friendly (e.g., by removing or re-siting wind turbines if they pose a relatively high collision risk for birds and bats, creating ecological corridors throughout solar farms).

4.3. Biodiversity and the location of electricity infrastructure

Various factors inform siting decisions, such as energy resource availability, technical feasibility for construction, connection to electricity grids, distance from settlements and environmental issues. As the potential impacts of renewable power on biodiversity are largely location-specific (see Chapter 3), decisions on where to deploy renewable power infrastructure are a critical entry point for mainstreaming biodiversity into energy planning. Explicit integration of biodiversity into siting decisions is fundamental for avoiding impacts on biodiversity, which is the first step of the mitigation hierarchy – an important tool to guide planning and projects (Box 4.4).

Key biodiversity factors to consider regarding the location of renewable power infrastructure include the location of areas legally established as sites of international, national, regional or local priority for biodiversity (e.g. Protected Areas; Natura 2000 sites in Europe; World Heritage sites); other areas identified as important for biodiversity through conservation planning prioritisation tools (e.g. Key Biodiversity Areas, Important Marine Mammal Areas, Wilderness Areas), the distribution of rare or threatened species and habitats, and migratory routes of birds, bats and other mammals.

Box 4.4. The Mitigation Hierarchy

The mitigation hierarchy provides a structured approach to development planning and the implementation of infrastructure projects with the aim of limiting the negative impacts on biodiversity. Implementation of the hierarchy is vital for achieving goals of no net loss of biodiversity or net biodiversity gain. The mitigation hierarchy involves four sequential and iterative steps. These are outlined below together with examples specific to renewable power:

1. **Avoidance:** Avoidance tends to be the most effective and least expensive way of reducing potential negative impacts on biodiversity. It requires biodiversity to be considered in the earliest stages of planning. Avoidance measures include, for example, careful siting of renewable power infrastructure to avoid important areas for biodiversity such as key biodiversity areas and migratory routes; or constructing infrastructure outside breeding seasons to avoid disturbance.
2. **Minimisation:** Minimisation measures reduce the duration, intensity and extent of impacts that cannot be completely avoided. It includes physical controls (e.g., bird flight diverters on power lines), operational (e.g., shut-down of wind turbines during migration) and abatement controls (e.g., technologies to reduce pile-driving noise for offshore wind).
3. **Onsite rehabilitation/restoration:** Impacts that cannot be completely avoided or minimised can be partly addressed through rehabilitation or restoration. Rehabilitation aims to return basic ecological functions and ecosystem services. Restoration aims to return an area to its original state.
4. **Offset:** Residual impacts may occur even after full implementation of the previous three steps. Offsetting aims to compensate for any residual adverse impacts through actions taken elsewhere to achieve positive biodiversity outcomes for the species or habitats affected by the development project. Through effectively designed offsets, developers could deliver no net loss or net gains in biodiversity.

Figure 4.1. Mitigation hierarchy

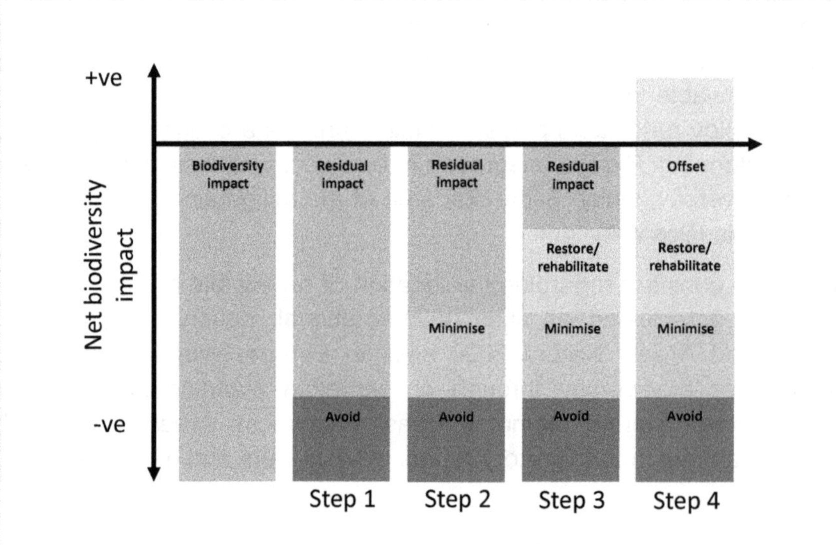

Source: Text based on (OECD, 2016[25]), Biodiversity Offsets: Effective Design and Implementation and (Bennun et al., 2021[26]), Mitigating biodiversity impacts associated with solar and wind energy development: guidelines for project developers. Figure adapted from (Rio Tinto, 2012[27]), "Rio Tinto and biodiversity: Working towards net positive impact", www.riotinto.com/documents/Rio_Tinto_and_biodiversity.pdf.

While some overlap between renewable power and areas of conservation importance is likely in the future, significant opportunities exist to deploy renewable power in areas of low-ecological risk (Dunnett et al., 2022[28]). Low-risk siting strategies include integrating electricity generation technologies and power lines with existing infrastructure (e.g. solar rooftops; for further discussion see 4.3.1) and deploying ground-mounted facilities in converted lands, as these tend to be lower in biodiversity than unconverted lands (Newbold et al., 2015[29]). Globally, sufficient converted land with renewable energy resource potential exists to deliver 17 times the required renewable energy to meet the combined emission reduction targets of the NDCs submitted by May 2016 (Baruch-Mordo et al., 2019[30]). In addition to avoiding and minimising biodiversity impacts, deploying renewable power infrastructure on converted lands may provide opportunities to deliver net gains for biodiversity if renewable power companies restore biodiversity on their facilities.

Brownfield sites are an example of converted land that may hold promise for low-impact siting. For example, a study of renewable power potential in West Virginia found more than 100 000 acres (>4 000 hectares) of former mine lands and other brownfields that were suitable for solar energy development. Many of these sites have existing infrastructure (e.g. road and power lines) and are close to energy markets, reducing the need for additional infrastructure expansion (James and Hansen, 2017[31]). A study of energy potential in low-ecological risk areas in Nevada concluded that developing solar on former minefields and brownfields could power 3.8 million homes (TNC, 2020[32]).

While targeting converted, unproductive land for renewable power development can reduce trade-offs between renewable power and biodiversity objectives, such areas may still have important biodiversity values that could be threatened by renewable power development. For example, in parts of Europe marginal agricultural lands provide critical habitats (Halada et al., 2011[33]). Deploying solar panels could negatively impact these habitats if it entails vegetation clearance or changes to the low-intensity agricultural practices shaping these habitats. Co-location of low-intensity agriculture with renewables could be one potential strategy for managing such cases (see 4.3.1). Similarly, some brownfield sites (e.g., old land-fill sites) support unique, relatively undisturbed ecological communities and contribute to landscape-level biodiversity (Macgregor et al., 2022[34]). Site evaluation and careful construction and operation therefore remains important for anticipating and addressing biodiversity impacts and other challenges (e.g. release of landfill gases if solar is developed on old landfill sites (AXA XL, 2022[35])).

Harnessing opportunities to develop renewable power on converted lands could require changes to land-use planning regulations and spatial plans. For example, to facilitate deployment of solar energy on former mine lands, the Administrative Code of Nevada was updated to explicitly list "renewable energy development and storage" as an acceptable post-production use for shuttered mining operations (Bream, 2018[36]). Deployment in degraded lands can also be encouraged through incentives (e.g., integration of biodiversity criteria into power procurement – see Chapter 5).

4.3.1. Multi-use spaces: co-locating renewable power with other activities

Opportunities exist to develop multi-use spaces or co-locate renewable power infrastructure. For example, renewable power infrastructure can be co-located with other power infrastructure (e.g., wind and solar; wind and storage), with other complementary infrastructure (e.g., roads, carparks, buildings) and economic activities (e.g., agriculture, aquaculture, forestry), and with biodiversity protection and restoration activities. Co-location can be an effective strategy for minimising land-use change (and therefore biodiversity risks) associated with renewable power expansion, while also harnessing synergies across different social, economic and environmental objectives.

Co-location of power infrastructure (e.g., wind turbines, solar PV/CSP, storage) is increasing across the globe, and considerable scope exists for further expansion. For example, the technical capacity of wind energy facilities in place in Australia to accommodate co-located solar farms was over 1 gigawatt (GW) in 2016, and further opportunities are emerging as renewable power expands (AECOM, 2016[37]). In India, at least 110 GW of wind and 360 GW of solar PV could be co-located, meeting 35% of India's electricity demand in 2030 (Deshmukh et al., 2019[38]). In addition to reducing land-use pressure, co-locating solar and wind power (or storage) can help smooth electricity generation (European Commission, 2020[39]). It can also provide private benefits. For example, it is estimated that co-location of solar and wind power in Australia could drive cost efficiencies of 3%-13% for capital expenditure and 3%-16% for operating expenditure, by amalgamating grid connections, development costs (e.g. ecological assessments) and equipment installation, among other things (AECOM, 2016[37]).

A second form of co-location is siting renewable power infrastructure with other infrastructure or economic activities. As mentioned above, one of the most effective approaches for avoiding biodiversity impacts from renewable power expansion is to integrate solar panels into the built environment. Rooftop solar is a well-known example, but solar panels are also starting to be integrated into other infrastructure such as parking lots and noise barriers along motorways (Rijkswaterstaat, 2020[40]). Such integrated approaches with solar panels deployed close to their end-use can also help reduce power losses associated with electricity transmission and distribution (van de Ven et al., 2021[41]).

For ground-mounted solar energy facilities, co-location depends on utilising the space below solar arrays or separating a plot of land into different uses (e.g. solar arrays could be deployed in the non-irrigated corners of centre-pivot irrigation plots) (Hassanpour Adeh, Selker and Higgins, 2018[42]). Co-location of solar energy and agriculture (agrivoltaics) is gaining increasing interest. Solar energy has been co-located with crops, small fruit trees, and small to medium-sized livestock.

Agrivoltaics have the potential to provide multiple additive and synergistic benefits. The main benefit for biodiversity arises from reduced land-use pressure (Amaducci, Yin and Colauzzi, 2018[43]). Other benefits include a diversified income stream for landowners, reduced plant drought stress, greater food production, increased PV panel efficiency due to cooling effect of vegetation under solar panels and reduced mowing requirements (Barron-Gafford et al., 2019[44]; Adeh et al., 2019[45]; Hassanpour Adeh, Selker and Higgins, 2018[42]). Locating solar on existing agricultural land may also increase community support, thereby facilitating permitting of renewable power projects (Pascaris et al., 2022[46]). Opportunities also exist to co-locate floating solar PV with aquaculture (Pringle, Handler and Pearce, 2017[47]).

Wind energy facilities often extend over large areas, due to the spacing requirements of turbines (see Chapter 3). However, only a small fraction of this area is occupied by infrastructure (turbine foundations, substation, access roads etc). The remaining space can be used for other activities. Onshore wind energy facilities, for example, are often combined with agriculture (crop and livestock systems) and increasingly forestry[6] (European Commission, 2020[39]). They may also be compatible with transport networks, industrial activities and public utility networks (RTE, 2022[2]).

A third form of co-location is combining biodiversity protection or restoration with renewable power. The deployment of renewable power infrastructure, whilst itself having potentially negative impacts on biodiversity, may reduce other pressures at the site. This could provide opportunities to combine energy generation with targeted restoration or protection measures. For example, recent studies demonstrate that through careful design, combining restoration activities and transmission infrastructure can increase biodiversity and provide a corridor for species to move across the landscape (e.g. to different foraging or breeding sites) (Ferrer et al., 2020[48]). Through strategic planting, restoration and maintenance, land-based solar PV facilities on degraded land could support pollinators, with potential spillover benefits for wild plants off-site and for agricultural crops (Hoffacker, Allen and Hernandez, 2017[49]; Randle-Boggis et al., 2020[50]; Blaydes et al., 2021[51]; Semeraro et al., 2018[52]; Sinha et al., 2018[53]; Dolezal, Torres and O'Neal, 2021[54]). Floating PVs could be designed to benefit aquatic plants and the fish that depend on them for shelter and food (de Rijk et al., 2021[55]).

In Europe, initiatives are underway to pilot artificial reefs at offshore wind farms, which are designed to support blue mussels, cephalopods, and flat oyster restoration (WaterProof Marine Consultancy and Services, 2020[56]; Didderen, Bergsma and Kamermans, n.d.[57]). To promote and guide efforts to co-locate wind energy, biodiversity restoration and food production, the Dutch Ministry of Agriculture, Nature and Food Quality commissioned guidelines on "Nature-Inclusive Design: A catalogue for offshore wind infrastructure" (Prusina et al., 2020[58]). The Nature and Environment Foundation *(Stichting Natuur & Milieu)* and North Sea Foundation *(Stichting De Noordzee)* commissioned a report on "Options for biodiversity enhancement in offshore wind farms Knowledge base for the implementation of the Rich North Sea Programme" (Waardenburg, 2020[59]).

Box 4.5. Co-locating solar and agriculture

Innovative Solar Practices Integrated with Rural Economies and Ecosystems (InSPIRE)

The U.S. Department of Energy's InSPIRE project brings together industry, academia and national laboratories to examine solar energy and environmental synergies across four core research areas: 1) Low-impact site preparation practices for ground mounted solar projects; 2) Improved compensatory mitigation planning; 3) Siting solar projects on contaminated and marginal lands; 4) Co-locating solar projects on agricultural lands for mutual benefits. The InSPIRE project combines innovative computational studies with field-based research to provide data to landowners, the solar industry and state decision-makers. It analyses native vegetation growth underneath and around ground-mounted solar installations; agricultural crop performance under innovative solar configurations; impacts of low-impact solar development approaches on soil quality, carbon storage, storm water management, microclimate conditions, and solar efficiencies; and benefits of pollinator-friendly solar on local agricultural yields.

Source: (Dreves, 2019[60]), Beneath Solar Panels, the Seeds of Opportunity Sprout; (DOE, 2022[61]), InSPIRE.

The viability and sustainability of co-location is context-specific (Ravi et al., 2016[62]). For example, establishing artificial reefs may be appropriate in locations where similar habitats once existed but have since been lost or degraded (e.g., from dredging), but inappropriate in areas where seagrass is the natural habitat. Furthermore, co-location can entail risks and trade-offs (AECOM, 2016[37]; Blyth-Skyrme, 2011[63]; Macknick, Beatty and Hill, 2013[64]), which will be important to understand and consider. For example, one risk of establishing artificial reefs at turbine foundations of offshore wind facilities is that invasive alien species benefit at the expense of local species (Degraer et al., 2020[65]). Combining renewable power generation and agriculture could in theory reduce the productivity of one or both activities relative to a single use siting approach, leading to partial displacement of these activities with potential biodiversity impacts (see indirect impacts in Chapter 3). Ongoing efforts to build the scientific and economic case for co-location, including through laboratory studies, pilot tests and monitoring, will be valuable for deepening the understanding of potential benefits and trade-offs of co-location in different contexts.

To capitalise on the potential benefits of co-locating activities, it may be necessary to adapt spatial planning, sectoral policy and infrastructure design. For example, the co-location of passive gear fishing with offshore wind farms may require, among other things, a definition of the legal basis, implementation of safety regulations and delineation of minimum requirements for fishing vessels (Stelzenmüller et al., 2016[66]). Co-locating solar energy in agricultural land may require designing elevated solar modules to optimise space for grazing or crop growing (RTE, 2022[2]). Effective implementation of co-location approaches will also benefit from the active engagement of local communities, economic interests and environmental experts (Schupp et al., 2021[67]), for example, by establishing communities of practice (Steins et al., 2021[68]).

4.3.2. Spatial planning and renewable energy zones

Well-designed and implemented spatial planning can facilitate a biodiversity-aligned transition to low emissions electricity systems. The importance of spatial planning in reconciling various demands on land and sea resources, including renewable power and biodiversity protection, is widely accepted. However, the extent and effectiveness to which spatial plans have been applied varies considerably across countries. Not all countries explicitly address biodiversity in their plans (Tucker, Quétier and Wende, 2020[69]). Good practice for spatial planning calls for an ecosystem approach[7] and a multi-sector strategy that balances and achieves environmental, economic and social objectives.

Governments can develop or adapt spatial plans to identify and prescribe appropriate locations for the development of renewable power infrastructure. Integrating biodiversity considerations into the spatial planning process could help avoid some of the significant adverse impacts of renewable power projects on biodiversity. For example, spatial planning could identify areas as low ecological risk, medium ecological risk or high ecological risk for renewable power development (European Commission, 2020[39]), drawing on tools such as sensitivity maps (see 4.4.1). This could then inform whether or under what conditions renewable power development should take place. Governments could exclude renewable power infrastructure from areas identified as critical for biodiversity or particularly vulnerable to renewable power developments. Alternatively, they could require projects in these areas to undertake more rigorous environmental assessments, mitigation measures and monitoring. Conversely, environmental permitting requirements could be less onerous at low-ecological risk sites.

In addition to steering development away from high-ecological-risk areas to low-ecological risk areas, renewable energy zones could deliver climate, energy and economic benefits. This is because the identification of low-risk areas for development may enable faster expansion of renewable energy with reduced project approval times and costs (see 5.1.1), while offering increased certainty and a level playing field for developers. For example, the U.S. Bureau of Land Management pre-approved low-ecological-risk renewable energy zones for solar energy development as part of the Department of the Interior's Western Solar Plan. Several large-scale solar projects were subsequently approved at these sites within 10 months, which is less than half the time that was usually required (US DOI, 2015[70]).

Spatial plans for renewable energy development can be developed at national, subnational and landscape level (see Box 4.6), and are applied in developed and developing countries (McKenney, 2020[71]). For example, EU Member States are required by EU Article 15(7) of the revised Renewable Energy Directive (2018/2001) to carry out an assessment of potential renewable energy sources and "where appropriate, include spatial analysis of areas suitable for low-ecological-risk deployment". In France, the national government asked regions to map areas favourable to wind energy development to help reach the objectives of the national multiannual energy programme, whilst accounting for biodiversity, landscape and human activity considerations. More than 35 biodiversity issues were considered in the mapping, which was carried out by elected representatives in the Regions, municipalities and the inter-municipalities and key stakeholders, such as environmental organisations. The regional mapping exercise is helping to inform a national map of appropriate sites (France, 2022[72]).

In the US, a process has been established for proactively identifying renewable energy zones (REZ) and planning the expansion of transmission infrastructure to connect these areas to the power grid. A REZ is defined in the US as an area that has high-quality renewable energy resources, suitable topography and land-use designations, and demonstrated interest from developers (Lee, Flores-Espino and Hurlbut, 2017[73]). Data on species, habitats, migratory routes and protected areas can be factored in when identifying a REZ (McKenney, 2020[71]). An Energy Zones Mapping Tool funded by the U.S. Department of Energy enables users to select from over 300 Geographic Information System data layers, including many related to biodiversity, to inform the REZ planning process. Several other countries, including India and 21 African nations have adopted similar approaches, many with the support of the United States Agency for International Development (USAID) (McKenney, 2020[71]).

Box 4.6. Planning for renewables and biodiversity at a landscape level: The Desert Conservation Plan

The Desert Renewable Energy Conservation Plan (DRECP) is a landscape-level plan for the development of utility-scale renewable power generation and transmission infrastructure in the deserts of Southern California. The plan covers 22.5 million acres (c. 9 million hectares) across seven counties. Its three objectives are to: i) advance federal and state national resource conservation goals and other management goals; ii) meet federal Endangered Species Act (ESA) and Federal Land Policy and Management Act (FLPMA) requirements; and iii) facilitate the timely and streamlined permitting of renewable power projects.

The plan was developed through a cross-sectoral collaboration of both federal and state agencies, together referred to as the Renewable Energy Action Team Agencies (REAT). REAT includes the Bureau of Land Management (BLM), U.S. Fish and Wildlife Service (USFWS), California Energy Commission (CEC), and California Department of Fish and Wildlife (CDFW). The planning process also involved consultation with other federal and state agencies, local agencies, tribal governments, private utilities and renewable energy proponents, and the public (citizens and not-for-profits). Furthermore, the DRECP was subject to a Programmatic Environmental Impact Assessment.

The DRCEP has three core components: i) a federal "general conservation plan" under the ESA, ii) a California "natural communities conservation plan", under California's Natural Communities Conservation Plan Act, and iii) a "Bureau of Land Management Land Use Plan Amendment (LUPA)", under the FLPMA. Central to this process, was the identification of locations most compatible with renewable energy development, understood to be areas with high renewable energy resource potential, close to existing or planned transmission and with relatively low biodiversity value.

The LUPA allows more than 800 000 acres (c. 324 000 ha) of land to be developed for renewable energy. Of this, 388 000 acres (157 000 ha) are designated as Development Focus Areas (DFAs), which are areas with substantial energy generation potential, access to existing or planned transmission, and low resource conflicts. Conservation and Management Actions were developed to provide certainty to help streamline and incentivise utility-scale renewable energy generation in DFAs. Approximately 40 000 acres (c. 16 000 ha) have been designated as Variance Process Lands (VPLs), where renewable energy development could be considered and approved without a plan amendment. Additionally, renewable energy could be developed in the 419 000 acres (c. 170 000 ha) of General Public Lands and 35 000 acres (c. 14 000 ha) of Extensive Recreation Management Areas (ERMAs) (not overlaid by a conservation allocation), but a plan amendment would be necessary as part of project review and approval. The LUPA identifies approximately 3 956 000 acres (c. 1 600 000 ha) of California Desert National Conservation Lands and allocates 6 527 000 acres (c. 2 641 000 ha) of total conservation designations for biological, cultural, and other natural resource protection and outlines managements actions for these areas.

Source: (US Bureau of Land Management, 2015[74]), Desert Renewable Energy Conservation Plan: Proposed Land Use Plan Amendment and Final Environmental Impacts Assessment. (US Bureau of Land Management, 2016[75]), Executive Summary for the Record Decision for the Land Use Plan Amendment to the California Desert Conservation Plan, Bishop Resource Management Plan, and Bakersfield Resource Management Plan.

Given the multiple economic pressures on marine resources and the rapid increase of offshore renewable energies, it is becoming increasingly important for spatial plans to cover not only terrestrial but also marine areas (OECD, 2017[76]). Responding to this need, the EU adopted a Maritime Spatial Planning (MSP) Directive 2014/89/EU that helps guide the sustainable development of marine-based energy technologies. The aim of the MSP Directive is to create a common framework for reducing conflicts and harnessing synergies across sectors such as maritime transport, fisheries and energy, to encourage cooperation and investment and preserve the environment (European Commission, 2020[39]). Members States were required to develop maritime spatial plans by 31 March 2021. An example of marine spatial planning for renewable energy in Germany is outlined in Box 4.7.

Box 4.7. Marine spatial planning in Germany

The marine spatial plan developed for Germany's exclusive economic zone (EEZ) in the North Sea aims to coordinate the growing conflict of maritime uses between space-intensive offshore wind farms and marine environmental protection goals, as well as traditional maritime uses such as shipping and fisheries. The Spatial Plan formulates guidelines for spatial development and sets targets and principles for the various maritime functions and uses. For the German EEZ in the North Sea the Spatial Plan contains provisions aimed at coordinating the individual uses and functions of shipping, the exploitation of resources, laying of pipelines and submarine cables, scientific marine research, wind power production, fisheries and mariculture, as well as protection of the marine environment. The spatial plan was subject to a strategic environmental assessment (SEA), in accordance with the SEA Directive 2001/42/EC. Broad public participation was secured through consultations with stakeholders from agencies and non-governmental organisations covering the following issues: marine environment and nature conservation, fisheries, energy, sand and gravel, shipping, military, tourism, leisure boating, research.

Source: (EC, 2022[77]), Maritime Spatial Plan for German EEZ in the North Sea.

It is good practice for planners to apply strategic environmental assessment (SEA) to the development of spatial plans (see Chapter 5). These should include an assessment of cumulative impacts on biodiversity, which are more easily assessed and addressed at a strategic planning level than at a project level. While spatial plans subject to an SEA may help avoid the worst of environmental impacts, they do not remove the need for environmental screening at the project level or even an environmental impact assessment. Rather, they provide a framework and initial siting prescriptions. Local specificities may not be fully captured in coarse grain spatial plans, and further opportunities for avoiding or minimising biodiversity impacts could arise through strategic micro-siting within renewable energy zones and other mitigation measures.

Spatial plans and land/sea-use restrictions may need to be periodically revised to reflect improvements in data and knowledge, evolution of renewable power technology and other social, economic and environmental factors. The Netherlands, for example, applies adaptive management to marine spatial planning (Vrees, 2021[78]). While adaptive management is a good practice, continuity and predictability are also important for businesses and investors. Adaptive management therefore ideally takes place within a broader framework shaped by a long-term vision (Vrees, 2021[78]). Ongoing monitoring and evaluation are fundamental to inform adaptive spatial planning (see 5.1.4).

4.4. Decision-support and planning tools

Planning for the low-emissions transition requires making informed decisions on what is best for society. This includes deciding on which technologies to adopt and where to locate electricity generation and transmission infrastructure. These are complex choices owing to the range of policy priorities (e.g., biodiversity protection, food provision, energy access), inevitable trade-offs and uncertainties that planners must deal with. Various tools exist to support these decisions, such as strategic environmental assessment (SEA), environmental (wildlife) sensitivity mapping, cost-benefit analysis (CBA) and multi-criteria decisions analysis (MCDA). Environmental sensitivity mapping and SEA are environmental assessment tools, while CBA and MCDA are standard economic appraisal tools, which can be tailored to account for biodiversity. These decision-support tools are complementary, not mutually exclusive. This chapter discusses environmental sensitivity mapping and biodiversity-inclusive CBA and MCDA. SEA is discussed together with EIA in Chapter 5 in the section on regulatory policy instruments.

4.4.1. Environmental sensitivity mapping

Environmental sensitivity mapping (or wildlife sensitivity mapping) can guide the deployment of renewable power infrastructure to help ensure it does not compromise biodiversity goals. It is a tool for identifying and communicating the location of biodiversity features (e.g. plant or animal species, habitats, ecosystems) that are vulnerable to renewable power developments because of their conservation status and their susceptibility to impacts (Bennun et al., 2021[26]). While different approaches exist, sensitivity maps typically draw on geographic information systems (GIS) and spatial data on species and habitats. Most approaches also assign sensitivity values to biodiversity components (Allinson et al., 2020[79]).

By predicting potential conflicts between renewable power development and vulnerable biodiversity features, sensitivity maps can be an important input for developing spatial plans during strategic planning and in initial project siting decisions. They can also be used as a due diligence and risk assessment tool for power purchasers and financial institutions investing in renewable power projects (Allinson et al., 2020[79]).

Various sensitivity mapping approaches have been developed, primarily by civil society or academic institutions, with varying degrees of government involvement. The maps typically operate at a landscape level, with regional, national or international coverage. Most approaches have been applied to onshore and offshore wind energy. For example, a 2020 review of sensitivity mapping identified 24 approaches or tools,[8] of which all but one had been developed primarily for wind power (Allinson et al., 2020[79]). Despite the focus on wind, sensitivity mapping can also be applied to other electricity generation infrastructure (see e.g. (Gove et al., 2016[21]) and transmission infrastructure (see e.g. (Gauld et al., 2022[80])).

Most sensitivity mapping approaches focus on birds, with only a handful covering bats or other mammals (Allinson et al., 2020[79]). Integrating other taxa and habitats into sensitivity maps could increase their value and uptake by decision makers. The focus on avian sensitivity has likely emerged for two reasons. First, birds are one of the most impacted groups by wind energy. Second, information and data on avian species distribution, abundance and risk factors are relatively advanced. Improving the underlying data on other taxa and the risks posed to them by renewable power infrastructure, will be fundamental for expanding the scope and use of sensitivity maps (see also 5.1.4).

Sensitivity mapping has been used to inform renewable power siting in several countries (Box 4.8). However, application of sensitivity mapping tends to be ad hoc rather than formalised in energy planning processes. Scope exists for governments to better integrate sensitivity mapping into their planning processes. This could involve developing guidance on sensitivity mapping (e.g. the EU commissioned a Wildlife Sensitivity Mapping Manual (Allinson et al., 2020[79])), leading the development of sensitivity maps (e.g. the Environmental Zoning Tool developed by the Spanish government Box 4.8), or establishing regulatory requirements to consider sensitivity maps as part of the planning and appraisal process. Ensuring the compatibility of sensitivity mapping with planning and consenting procedures is also important, which underscores the value of government engagement in tool development. For example, as biodiversity impacts are just one constraint to consider when siting infrastructure, governments could find value in wildlife sensitivity maps that can be integrated with information on other relevant constraints or criteria, such as resource potential, land-uses, legal restrictions and distance from the transmission grid. Finally, ongoing efforts to improve the coverage of species and habitats of concern will help to increase sensitivity mapping's utility.

Box 4.8. Sensitivity mapping examples

Environmental Zoning for Implementing Renewable Energy: Wind and Solar – Spain

Spain's Ministry for the Ecological Transition and Demographic Challenge developed a sensitivity mapping tool to guide the development of onshore wind and solar PV. The tool is intended to be indicative, providing strategic guidance for project siting. It complements regulatory requirements. The tool assesses sites against 18 categories of indicators covering e.g., ecological corridors, Important Bird and Biodiversity Areas, Natura 2000 sites, Ramsar wetlands of international importance, UNESCO sites, visual impacts etc. Some indicators are exclusionary (i.e., development is not advised if a certain feature is present). The other indicator scores are weighted to produce an overall environmental sensitivity score for the area in question. The zoning tool, available on the Ministry's website, consists of two layers of information (one for energy and one for wind), with colours indicating the sensitivity score of any point on the map of Spain.

Sensitivity Mapping for Mammalian Carnivores – Croatia

Disturbance and habitat fragmentation are key threats to Croatia's carnivores. Therefore, poorly planned wind energy developments could threaten their survival in the region. The University of Zagreb and the Croatian Ministry for Environment and Nature created a sensitivity map in 2016 based on data on large carnivore occurrence and related habitat characteristics. The sensitivity map defines nine classes of habitat sensitivity, which are grouped into four categories of suitability for large mammal presence. Individual sensitivity maps were developed for bears, wolves and lynx. Two additional maps were produced for bear and wolf reproduction sites. Finally, one combined sensitivity map was created for all three species together. The sensitivity mapping is accompanied with guidelines and information on the habitat use of carnivores. The analysis and guidelines can be used to evaluate and inform all environmental assessment procedures, such as SEAs and EIAs. The analysis and guidelines include advice about which areas to prioritise for avoidance and estimations of the permissible losses of each habitat class where some loss is unavoidable.

Soaring Bird Sensitivity Mapping Tool – Southern Europe, Middle East and North Africa

BirdLife International's Soaring Bird Sensitivity Mapping Tool provides public access to extensive spatial datasets relating to soaring birds. A formula assigns sensitivity categories to allow an objective assessment and comparison of prospective locations based on available data. The aim of the tool is to provide planning authorities, developers and other stakeholders with an authoritative, transparent and accurate assessment of the soaring bird sensitivity of onshore wind energy and transmission infrastructure. The tool is based on data for 89 species, primarily from BirdLife's Important Bird and Biodiversity Areas (IBA) dataset. Supplementary information, including soaring bird satellite tracking data and species' range maps, as well as spatial data on protected areas and relevant topography, are also included to provide additional context and insight.

Site Renewables Right – United States

The Nature Conservancy's Site Renewables Right tool was designed to inform the siting of wind and solar energy projects in the Central United States (17 states) where approximately 75% of US onshore renewable energy is likely to be located by 2050. Site Renewables Right is a sensitivity map for buyers, developers, planners and policy makers to identify areas where renewable energy development is unlikely to lead to wildlife conflicts, project delays or related cost overruns. It comprises two distinct layers, one for wind and one for solar energy. The maps are suited to landscape-level site evaluations and site characterization analyses. They are not intended for use as "go/no-go" maps, nor as a substitute for robust assessment of anticipated impacts to species and habitat. Sensitive areas covered in the maps include: whooping crane stopover sites with 400m buffer (wind and solar); big game habitats (wind); eagle and other raptor nesting areas (wind); water, wetlands, and riparian corridors (wind and solar); breeding waterfowl habitats (wind); protected and managed lands (wind and solar); important bird areas (wind); intact natural habitats (wind and solar); bat roosts (wind); other areas of biodiversity significance (wind and solar); threatened and endangered species (wind and solar); climate-resilient lands (wind and solar).

Source: (Passoni et al., 2017[81]), Framework for strategic wind farm site prioritisation based on modelled wolf reproduction habitat in Croatia, 10.1007/s10344-017-1092-7; (Allinson et al., 2020[79]), The Wildlife Sensitivity Mapping Manual: Practical guidance for planning in the European Union. Final report for the European Commission (DG ENV), https://ec.europa.eu/environment/nature/natura2000/management/nat; (Spain, 2020[82]), Environmental zoning for : Wind and solar PV www.miteco.gob.es/es/calidad-y-evaluacion-ambiental/temas/evaluacion-ambiental/zonificacion_ambiental_energias_renovables.aspx; (TNC, 2022[83]), Site Renewables Right: Accelerating a Clean and Green Buildout in the Central United States; www.nature.org/content/dam/tnc/nature/en/documents/SRR_Methods_20220202_LR.pdf.

4.4.2. Cost-benefit analysis and multi-criteria decision analysis

Cost-benefit analysis (CBA) and multi-criteria decision analysis (MCDA) are two related decision-support tools that are sometimes used to inform decisions on electricity generation portfolios and power sector policies and projects. These tools can be used individually, in parallel (as two separate sources of information), or in an integrated manner where monetised and non-monetised values are considered together (e.g., CBA results can be one criterion in the MCDA). Both tools are increasingly combined with GIS to inform spatially explicit decisions, such as infrastructure siting.

Effective integration of biodiversity into CBA and MCDA, when appraising power sector policies, plans, programmes (hereafter "policies") and projects, can help ensure that choices made in power-sector planning optimally balance trade-offs across biodiversity and other environmental and social objectives. This requires thorough accounting of the impacts on ecosystem services in CBA and inclusion of explicit and appropriately weighted biodiversity criteria in MCDA. As illustrated in the examples below, investment options could be pre-screened with exclusionary criteria for those that may reduce the viability of species or ecosystems of conservation concern.

This section provides a brief overview of CBA and MCDA and explores how biodiversity has been integrated into these tools in theory and practice, with a particular focus on renewable power development.

Cost-benefit analysis

CBA is a widely used economic appraisal tool that compares the social costs (reductions in human well-being) and benefits (increases in human well-being) of plans, policies or projects (Beria, Maltese and Mariotti, 2012[84]; OECD, 2018[85]). The benefits and costs are aggregated and monetised, accounting for different points in time. The result of a cost-benefit analysis is generally represented in a Benefit-Cost Ratio (BCR) or Net Present Value (NPV). For a policy or project to qualify on cost-benefit grounds, its social benefits (represented in monetary terms) must exceed its social costs (OECD, 2018[85]). Environmental cost-benefit analysis is the use of CBA to evaluate an environmental project or a project that will have a non-negligible impact on the environment.

Generally, CBA analyses in the context of renewable power did not account for the social costs and benefits associated with biodiversity and ecosystem services (but have rather focused on greenhouse gas emissions or air pollution). (OECD, 2018[85]). However, valuing ecosystem services – and incorporating ecosystem service values into CBA – is an important component of robust decision making. While challenges remain in valuing ecosystem services, a range of valuation approaches have emerged and substantial progress has been made in assigning increasingly robust values to ecosystem services even where these are not explicit in market prices (see section 13 of (OECD, 2018[85]) for an overview). These advances enable many ecosystem service values to be explicitly integrated into CBA. National appraisal frameworks in various countries, such as Australia (Infrastructure Australia, 2021[86]) and the UK (HM Treasury, 2022[87]) explicitly promote or facilitate inclusion of monetised ecosystem service values when conducting CBA. Other appraisal frameworks promote the integration of biodiversity (and other difficult-to-monetise impacts) in CBA using non-monetised values (e.g., through plus-minus methods (Norway Directorate for Financial Management, 2018[88]) – see discussion below on MCDA and integrated approaches).

Where decision makers choose to use CBA to inform electricity system design and technology choices or appraise renewable energy infrastructure projects,[9] integrating ecosystem service values in CBA provides an entry point for mainstreaming biodiversity in power sector planning. Take the example of transmission planning in the US. A key step in the transmission planning process outlined by USAID and the National Renewable Energy Laboratory guidelines (Lee, Flores-Espino and Hurlbut, 2017[73]), is to compare transmission options to understand the social costs, benefits and trade-offs of each option. For example, planners may need to choose between developing renewable power facilities at less productive sites that are close to existing transmission infrastructure or at more productive sites that require significant expansion of transmission infrastructure. The guidelines suggest CBA be used to compare these options and determine which approach has the greatest societal benefit. Ideally, such an analysis would account for the biodiversity impacts of each of these options, and how these translate to a change in welfare.

Examples of where biodiversity has been integrated into CBA for renewable power planning and project appraisal are not common in the literature and are mainly theoretical rather than applied. While various examples exist in academic literature of environmental CBA or economic valuation for evaluating renewable energy portfolios, policies or projects (Rhodes et al., 2017[89]; Pojadas and Abundo, 2022[90]; Rouhani et al., 2016[91]), these tend to focus on greenhouse gas emissions and air pollution rather than biodiversity. Those CBA studies that do consider biodiversity often do so by excluding ecologically important areas before or after CBA is applied, rather than attempting to value changes in ecosystem services (Deshmukh et al., 2019[38]; Sun et al., 2013[92]; Kim, Jang and Kim, 2018[93]).

One example of where CBA approaches have been applied to biodiversity and renewable power is an analysis of external costs and benefits of electricity generation options (hydropower, offshore and onshore wind, biomass) in Scotland (Bergmann, Hanley and Wright, 2006[94]). Using choice experiments the analysis showed that projects can have varying costs in terms of landscape, wildlife and air pollution impacts. They concluded that wildlife was highly valued by the population and that projects potentially harming wildlife would therefore require large offsetting benefits. A separate study (Álvarez-Farizo and Hanley, 2002[95]) used contingent rating and choice experiments to identify the social costs of environmental impacts of a wind farm in Spain. The analysis concluded that flora and fauna impacts were valued more highly than impacts on landscape and geologically rare cliff sites, but all three variables were significant determinants of preferences for wind power investments. The authors propose that such methodologies would facilitate wind farm development that minimises total social costs and maximises net benefits. A third study, in the US, applies CBA to compare costs and benefits of pollinator-friendly solar PV, standard solar PV and agriculture land, and provides policy recommendations based on their findings (Box 4.9).

Environmental CBA has well-known limitations and challenges (Beria, Maltese and Mariotti, 2012[84]; OECD, 2018[85]). These include its data and resource intensiveness, difficulties in accurately capturing all values and the uncertainty of ecosystem thresholds or tipping points. Nonetheless, it can play an important role as part of a broader policy or project appraisal process (Beria, Maltese and Mariotti, 2012[84]; OECD, 2018[85]). It is likely most useful when it is used as a decision-support tool, rather than as a prescriptive tool (Turner, 2007[96]). While it is unrealistic to include all ecosystem service values in CBA, ongoing improvements in ecosystem service valuation facilitate increasing integration of biodiversity considerations into CBA to ensure it is not ignored. Evaluating non-monetary criterion next to monetary criteria – including through integrated CBA-MCDA (discussed below) – can enable a more comprehensive and transparent assessment of positive and negative impacts in policy and project appraisal.

Box 4.9. Policy interventions for pollinator-friendly solar

Researchers at Yale University conducted a CBA to compare private and social returns from three types of land use: conventional solar, pollinator-friendly solar and pre-existing agricultural land uses. The analysis monetised positive and negative environmental externalities related to carbon emissions, soil erosion, groundwater recharge and pollination. It calculated the private and social rents from each land use, discounting values to their net present value for a 30-year project lifetime using a social discount rate of 3%. Total welfare gains and losses were calculated by comparing net gain in private, social and tax benefits across all land uses. The analysis revealed significantly higher private and social net benefits for conventional solar than farmland, and even higher net benefits for pollinator-friendly solar. Private benefits from pollinator-friendly solar projects were higher than from conventional solar because of the assumed efficiency gains for pollinator-friendly solar panels due to ambient microclimate conditions created by native plantings, and from lower operations and maintenance costs owing to reduced mowing for native plants compared to turfgrass. The relatively high social benefits associated with pollinator-friendly solar arose from greater groundwater recharge, reduced soil erosion and increased crop yields resulting from pollination (when projects are sited near pollinator-dependent farmland). The report outlined the following policy implications:

1. Private benefits: If investment does not increase in pollinator-friendly solar despite the assumed private benefits, information or behavioural market failures may be present. To address information gaps, research and development funding could focus on the relationship between native plantings, microclimates and solar panel output. If ongoing research confirms efficiency gains but investment does not increase, a behavioural failure may persist. Addressing behavioural barriers requires other measures, such as webinars or local permitting requirements that familiarise developers with pollinator-friendly practices. Additionally, stakeholder engagement across potential collaborating organisations can cultivate consensus and awareness.
2. Social benefits: The additional social (environmental) benefits of pollinator-friendly solar compared to conventional solar and agriculture warrant policy intervention. These benefits could best be captured by providing developers with a power purchase agreement "adder" that represents the per-kWh social benefits of pollinator-friendly solar. Second-best policy measures include a pollinator-friendly solar mandate, or tax credits to subsidise the practice. Research and development funding could improve ecosystem valuation and inform the appropriate incentive amount to efficiently internalise the ecosystem service benefits.
3. Location-specific benefits: Ecosystem service benefits are spatially specific. Values for soil carbon sequestration and groundwater recharge are highly dependent on topographies and soil types. Pollinator-friendly solar benefits depend greatly on the surrounding crop types. Policies need to be designed to direct and reward the siting of pollinator-friendly solar projects on lands where they will realise maximum social benefit. For example, policies could encourage project development near crops that depend on pollinators. The analysis modelled a 6.3% crop yield increase for a moderately pollinator-dependent crop (soy) from pollinator-friendly solar installations, but the yield increase for specialty crops could be much greater. Site-specific policy examples include property tax abatement, streamlined permitting processes and a tiered subsidy system where the subsidy is scaled up or down based on the value of ecosystem benefits that a project helps realise in the area where it is developed.

Source: (Siegner et al., 2019[97]), Maximizing land-use benefits from utility scale solar: A cost-benefit analysis of pollinator-friendly solar in Minnesota, https://cbey.yale.edu/sites/default/files/2019-12/MaximizingLandUseBenefitsFromUtility-ScaleSolar_0.pdf.

Multi-criteria decision analysis

Multi-criteria decision analysis (MCDA) is an approach for ranking or selecting alternative policies or projects. An advantage of this approach is its ability to account for a diverse range of social, economic and environmental dimensions – including those not easily monetised – in a single framework (Beria, Maltese and Mariotti, 2012[84]; OECD, 2018[85]). It also tends to be more "transparent" than CBA, since the objectives and criteria are usually clearly stated, rather than assumed (OECD, 2018[85]). On the other hand, unlike CBA, MCDA only provides a relative ranking of options; it does not determine whether it is appropriate to adopt a policy or project. Additionally, it is often unclear how MCDA deals with time discounting, which is a key part of CBA. Furthermore, as MCDA proceeds by adopting scores and weights chosen by experts, it tends not to be as "accountable" as CBA where the money units reflect individuals' preferences rather than expert preferences (OECD, 2018[85]). The capacity of MCDA to articulate ecosystem service values, its transparency and accountability, depend in part on the methods applied and how the process is organised and facilitated (Saarikoski et al., 2016[98]).

Typical steps for MCDA include establishing the goals or objectives of the policy or project, selecting criteria or attributes to achieve the objectives (measured in monetary or non-monetary terms), and then scoring and weighting the options for achieving the objectives. The outcome, in the simplest form of MCDA, is a weighted average of the scores with the option with the highest weighted score being the best (OECD, 2018[85]).

MCDA can help decision makers to select among electricity generation portfolios based on a range of criteria. It allows biodiversity impacts to be considered alongside other environmental, social and economic criteria such as greenhouse gas emissions and levelised-cost. For example, (Nock and Baker, 2019[99]) applied MCDA to evaluate the sustainability of electric generation portfolios in New England against seven sustainability criteria, one of which was land-use. A similar approach was taken to compare the sustainability of 13 individual electricity generation technologies in the US (Klein and Whalley, 2015[100]). However, the authors note that their land use results underscore the need for more research to harmonise land use estimates across energy options and reduce uncertainty, especially for life cycle estimates.

MCDA integrated with global information system (GIS) has been applied in studies to inform the siting of solar and wind energy facilities in various geographies (Shorabeh et al., 2019[101]). Biodiversity considerations have been addressed in GIS-MCDA in two ways: as a constraint and as criteria. Several studies on renewable energy siting apply biodiversity constraints, but then do not explicitly include biodiversity as criteria (Shorabeh et al., 2019[101]; Sánchez-Lozano, García-Cascales and Lamata, 2016[102]). For example, (Shorabeh et al., 2019[101]) applied several biodiversity constraints, including "must not be in a conservation area" and "must not be in wetlands, forests, agricultural areas", before applying ten criteria related to cost (e.g. distance from road, slope) and benefit (e.g. solar radiation, land surface temperature).

Other GIS-MCDA studies explicitly incorporate biodiversity as a criterium or both as a constraint and a criterium. For example, a GIS-MCDA for Gwadar, Pakistan, examined seven evaluation criteria, including two focused on biodiversity – avian hotspots and location of wetlands and forests (Zahid et al., 2021[103]). In their analysis of suitable locations for wind and solar developments in southern England, (Watson and Hudson, 2015[104]) created a binary constraint layer identifying entirely unsuitable locations, including "wildlife designations".[10] Seven criteria were then considered covering technical (wind speed, solar radiation), visual (distance from historically important sites, distance from residential areas), ecological (distance from wildlife designations), and economic (distance from transport links, distance from network connection) factors. Based on these criteria, locations were scored as not suitable, least suitable, moderately suitable and most suitable. Similarly, an assessment of potential wind energy locations in the Städteregion Aachen in Germany (Höfer et al., 2014[105]) first excluded natural resources areas, national parks, important habitat areas, bird reserves and protected biotopes. It also excluded a 300m buffer zone around these areas. After the exclusions zones had been removed, the criteria used to select sites included "distance from natural environments" and "land cover type".

Owing to the challenges of fully accounting for biodiversity and other complex societal values in CBA, it is increasingly common to complement CBA with MCDA. Integrated CBA-MCDA methodologies are also emerging (e.g. balance sheet approach; plus-minus methods) (Saarikoski et al., 2016[98]; Turner, 2016[106]; Norway Directorate for Financial Management, 2018[88]). ENTSO-E's Draft Cost Benefit Analysis of Grid Development Projects Guideline[11] (ENTSO-E, 2019[107]), for example, combines qualitative, quantitative non-monetised and monetised assessments, recognising that a fully monetised approach is not practically feasible. When impacts on biodiversity are predicted, the estimated impact mitigation costs (i.e., private costs to business) are captured in the CAPEX. Residual impacts on biodiversity (i.e., costs to society or externalities) are assessed through a quantitative but non-monetary indicator: the number of kilometres of overhead line or underground/submarine cable passing through sensitive areas.[12] A similar approach using kilometres of submarine cable as a proxy for biodiversity impacts was taken to communicate the importance of planning offshore transmission in the US (Pfeifenberger et al., 2023[108]).

4.5. Institutional co-ordination

Through its demand for land, renewable power expansion has implications for various policy areas, including biodiversity, climate change and production sectors (e.g., agriculture and fisheries). Effective co-ordination across the institutions that are responsible for these policy areas and house relevant expertise (i.e., horizontal co-ordination) is therefore critical for policy alignment.

The appropriate structure for facilitating horizontal co-ordination is likely to be country- and context-specific, but could involve e.g., inter-ministerial or inter-agency committees and working groups, taskforces or consultation processes that promote multi-stakeholder dialogue and inclusive decision-making processes. Coordination may be *ad hoc* or institutionalised. Examples are highlighted below:

- In Egypt, to help minimise conflicts between wind power development and migratory bird conservation, the Egyptian Environmental Affairs Agency (EEAA) and the New and Renewable Energy Authority (NREA) signed a Memorandum of Understanding (MoU), in 2012. The MoU provided a framework for co-operation on the sustainable deployment of renewable energy. In 2015, the Regional Center for Renewable Energy and Energy Efficiency initiated an Active Turbine Management Project (ATMP) based on a protocol signed with the EEAA, the NREA and the Egyptian Electricity Transmission Company. The ATMP is divided into 3 main sub-programmes: i) a bird monitoring programme; ii) a shutdown on demand programme; and iii) a fatality monitoring programme. These programmes are pertinent to all wind facilities in the Gulf of Suez region (RCREEE, 2015[109]).
- In France, until 2022, the responsibility for biodiversity, climate and energy policy was under the same ministry – the Ministry for Ecological Transition. Within the Ministry, a staff member in the directorate for water and biodiversity was responsible for following offshore wind development, while within the energy directorate some staff focused on biodiversity. A working group established in 2018 to examine cumulative effects of offshore wind was co-chaired by staff from each of these two directorates. The working group consisted of 25 central and local administrative representatives, academic experts and environmental protection agencies (France, 2022[72]).
- The Norwegian Water Resources and Energy Directorate (NVE) collaborated with the Norwegian Environment Agency, and another ten institutions to develop a knowledge base presenting positive and negative effects of wind power and information about regulatory responsibilities and procedures (NVE, 2022[110]).

Previous OECD work on biodiversity mainstreaming and policy alignment (OECD, 2018[111]; OECD, 2020[112]) and country responses to OECD's questionnaire on biodiversity and renewable energy highlight key lessons for institutional coordination. These include the importance of setting clear goals for coordinating bodies, establishing roles and responsibilities, and ensuring continuity of effort to facilitate progress towards the set goals. Ensuring sufficient financial, technical and time resources is fundamental.

In addition to horizontal coordination, delivering on renewable energy and biodiversity objectives requires vertical coordination to align national and subnational policy, and ensure national plans reflect local interests and issues. The energy transition is planned and delivered across multiple tiers of government (international, regional, national and local) (Goldthau, 2014[113]). While the share of decision-making power and responsibility across national and subnational institutions differs from one country to another, subnational governments are playing an increasingly important role in the energy transition. Generally, renewable power objectives are established at a national and international level, while subnational governments hold responsibility for land-use decisions and the implementation of renewable energy policies (Koelman, Hartmann and Spit, 2021[114]). As renewable power constraints and biodiversity impacts are highly location-specific, subnational institutions and stakeholders are well-positioned to inform national planning and ensure that the implementation of energy plans is biodiversity-aligned.

Disconnects or tensions between international, national, regional and local levels in renewable power planning can undermine efforts to sustainably scale up renewable energy (Michalena and Hills, 2012[115]). For example, a study of case studies on wind power development in the Netherlands found conflicting differences between local interests (liveability), regional interests (landscape protection) and national tier interests (renewable energy objectives) (Koelman, Hartmann and Spit, 2021[114]). Indeed, it is often subnational government institutions implementing national climate-energy policy that are confronted by resistance from individual landowners or communities, owing to the perceived impacts of renewable energy developments on ecosystem services (e.g. cultural services [aesthetic values]) (Zoellner, Schweizer-Ries and Wemheuer, 2008[116]; Larsson and Emmelin, 2015[117]).

Establishing processes or structures for vertical coordination can help to align interests across the different tiers of government and ensure policy coherence. As with horizontal coordination, it is important to clearly define the responsibilities of institutions across the different tiers of government, and to ensure that each institution has the capacity to carry out its responsibility (Goldthau, 2014[113]).

4.6. Cross-border collaboration

Collaboration across national borders or subnational borders can help to ensure the global transition to low-emissions electricity systems does not compromise efforts to halt and reverse biodiversity loss. First, the ranges of species and ecosystems affected by renewable energy development – particularly migratory birds, bats and marine species – extend beyond subnational and national borders. The biodiversity impacts of renewable energy projects in multiple jurisdictions can therefore accumulate, potentially leading to population or ecosystem declines (see Chapter 3). Collective ambition by sub-national and national governments to mainstream biodiversity into renewable energy development is necessary to ensure a species or ecosystem is protected across its entire lifecycle and range. It can also help avoid leakage where renewable energy projects shift towards jurisdictions with fewer regulatory requirements.[13]

Cross-border spatial planning (e.g., at a landscape or seascape level), coordinated monitoring of affected populations and data-sharing can also play an important role in managing cumulative impacts. The EU, for example, encourages and provides guidance for its member states on cross-border co-operation in marine spatial planning. At the global level, the Convention on Migratory Species provides a framework for co-operation to address negative impacts on migratory species, including those arising from renewable power infrastructure.

Second, cross-border trade in electricity presents both an opportunity and risk to biodiversity. Well-planned interconnection of grids across country (or state) borders can reduce system costs, balance supply and demand and increase energy access (European Commission, 2020[39]). Furthermore, by offering greater flexibility for siting, cross-border grid expansion could in some contexts increase opportunities for siting infrastructure in areas of low ecological risk and reduce overall need for electricity generating assets. The Californian study highlighted in Box 4.3, for example, concluded that if California were to access renewable resources from Western states, both biodiversity impacts and portfolio costs would decrease compared to a business-as-usual scenario for expansion (Wu et al., 2019[1]).

The risk of electricity trade is that it drives land-use pressure and biodiversity loss far from where the electricity is consumed (Holland et al., 2019[118]). Negative impacts could arise from the construction and operation of utility-scale wind and solar facilities (in the absence of robust planning and policy), or from the unsustainable harvest and export of biomass to fuel power plants. Furthermore, interconnecting grids entails greater extensions of electricity transmission infrastructure, which could negatively impact biodiversity for example through habitat fragmentation and by posing collision risk for volant species. This further emphasises the importance of ensuring all jurisdictions have effective policies and planning processes in place to safeguard biodiversity, and the need to ensure that cross-border energy trade is governed by good practice principles for safeguarding biodiversity. The Green Grids Initiative – One Sun One World One Grid, launched at the COP26 World leaders Summit by the UK and Indian Prime Ministers and backed by over 80 countries, could play a role in promoting grid interconnections that protect nature. In addition to harnessing synergies and reducing trade-offs across climate, energy and biodiversity goals in electricity trade, it will be important to address trade-related risks associated with sourcing of critical minerals (Box 4.10).

Box 4.10. Trade-related risks to biodiversity from critical minerals in renewable energy

Trade-related risks to biodiversity arise from the sourcing of critical minerals upon which renewable energy infrastructure depends. These minerals are globally unevenly spread (see Chapter 3), and in most cases will be sourced through international supply chains. International co-operation is fundamental to ensure that the extraction and trade of minerals is conducted sustainably, for example through application of appropriate standards and due diligence guidelines and increased supply chain transparency. The OECD, for example, has established Due Diligence Guidance for Responsible Supply Chain of Minerals from Conflict-Affected and High-Risk Areas and a Practical Tool on Environmental Due Diligence in Mineral Supply Chains.

Notes: OECD's Due Diligence Guidance is available here: (OECD, 2023[119]), OECD Practical Tool on Environmental Due Diligence In Mineral Supply Chains,
www.umweltbundesamt.de/sites/default/files/medien/6232/dokumente/211102_factsheet_practicaltoolenvddmineralsc.pdf.
Source: Author.

Third, development co-operation plays an important role in mainstreaming biodiversity (OECD, 2018[111]; Drutschinin et al., 2015[120]). Through technical and financial support, official development finance could support partner countries to deliver a biodiversity-aligned transition to low-emissions electricity systems. For example, Official Development Assistance (ODA) could be used to strengthen the evidence and knowledge base on renewable energy impacts on biodiversity given this is relatively poor in many developing countries (see Chapter 3), develop individual, institutional and systemic capacities to mainstream biodiversity into climate and energy planning and policy, support the development of sensitivity mapping tools or biodiversity-explicit spatial plans for renewable power, enhance monitoring frameworks and data management systems, and develop guidelines on biodiversity-inclusive SEA and environmental impact assessment (EIA), including cumulative impact assessments.

ODA from members of the OECD Development Assistance Committee has been promoting such biodiversity-related capacity development type of activities in the energy sector, to the tune of USD 2.4 million on average per year over 2011-20. This represents 1% of what these donors spent on biodiversity-related energy ODA over that period. Although overall biodiversity-related criteria in ODA funded energy activities have increased over this period, the share of commitments for biodiversity-related capacity development in the energy sector is still low, thus indicating the need for donors to continue integrating biodiversity considerations through their technical and knowledge transfer support.

Fourth, as knowledge of impacts and mitigation measures is rapidly evolving but incomplete, value lies in international exchange of knowledge, insights and experience. This could involve, for example, sharing information and evidence on species risk and effective mitigation measures, and exchanging good practices and tools for assessing species risk and environmental impacts. Multi-country initiatives have been established to this end (Box 4.11). Donors are also actively engaging in such exchanges through the promotion of triangular co-operation, although this is a development co-operation modality that has scope to grow in the area of biodiversity (OECD, 2019[121]), especially beyond Latin America and the Caribbean (OECD, 2023[122]).

Box 4.11. International initiatives to share knowledge on renewable energy and biodiversity impacts

The Convention on Migratory Species' Energy Task Force

The CMS Energy Task Force brings together governments, multilateral environmental agreements, investors, the private sector and non-governmental organisations with an aim of avoiding and minimising negative impacts of energy developments on migratory species. It was established in 2015 in accordance with Resolution 11.27 (Rev. COP13) Renewable Energy and Migratory Species. It was established in 2015 in accordance with Resolution 11.27 (Rev. COP13) Renewable Energy and Migratory Species to support the implementation of these resolutions and the use of relevant guidelines. The Task Force works to:

- Promote and develop guidance and tools for the sustainable deployment or retrofitting of renewable energy technologies and power lines.
- Exchange and disseminate good practices for deploying energy infrastructure.
- Provide recommendations on how best to respond to specific problems.
- Deliver research to address knowledge gaps.

WREN (Working Together to Resolve Environmental Effects of Wind Energy)

WREN, also known as Task 34, was established by the International Energy Agency (IEA) Wind Committee in October 2012 to address environmental issues associated with commercial development of land based and offshore wind energy projects. As the operating agent for WREN, the United States (US) leads this effort with support from the National Renewable Energy Laboratory (NREL), the Pacific Northwest National Laboratory (PNNL), and the US Department of Energy's (DOE) Wind Energy Technologies Office (WETO). The primary objective of WREN is to facilitate international collaboration and advance global understanding of potential environmental effects of wind energy. As the operating agent for WREN, the United States (US) leads this effort with support from the National Renewable Energy Laboratory (NREL), the Pacific Northwest National Laboratory (PNNL), and the US Department of Energy's (DOE) Wind Energy Technologies Office (WETO). To support this effort, *Tethys* (a US initiative developed in 2009) was expanded to serve as a collaborative outreach and engagement space, and to disseminate knowledge and information. Through WREN and Tethys, interested stakeholders can locate the latest scientific publications and briefings on the environmental effects of wind, and how to assess and mitigate them.

Source: (CMS, 2022[123]), CMS Energy Task Force, www.cms.int/en/taskforce/energy-task-force; (Hein, Whiting and Page, 2022[124]), About WREN, https://tethys.pnnl.gov/about-wren.

References

Adeh, E. et al. (2019), "Solar PV Power Potential is Greatest Over Croplands", *Scientific Reports*, Vol. 9/1, https://doi.org/10.1038/s41598-019-47803-3. [45]

AECOM (2016), *Infrastructure Interdependencies and climate risks report*, https://arena.gov.au/assets/2016/01/AECOM-Wind-solar-Co-location-Study-1.pdf (accessed on 10 March 2018). [37]

Allinson, T. et al. (2020), *The Wildlife Sensitivity Mapping Manual: Practical guidance for rewnewable energy planning in the European Union. Final report for the European Commission (DG ENV)*, https://ec.europa.eu/environment/nature/natura2000/management/nat. [79]

Álvarez-Farizo, B. and N. Hanley (2002), "Using conjoint analysis to quantify public preferences over the environmental impacts of wind farms. An example from Spain", *Energy Policy*, Vol. 30/2, pp. 107-116, https://doi.org/10.1016/s0301-4215(01)00063-5. [95]

Amaducci, S., X. Yin and M. Colauzzi (2018), "Agrivoltaic systems to optimise land use for electric energy production", *Applied Energy*, Vol. 220, pp. 545-561, https://doi.org/10.1016/j.apenergy.2018.03.081. [43]

AXA XL (2022), *Solar and wind power: Renewable Energy Environmental Exposures*, AXA, https://axaxl.com/-/media/axaxl/files/pdfs/fff/2022/6874_env_solar-wind-energy-exposures_whitepaper.pdf (accessed on 25 July 2022). [35]

Barron-Gafford, G. et al. (2019), "Agrivoltaics provide mutual benefits across the food–energy–water nexus in drylands", *Nature Sustainability*, Vol. 2/9, pp. 848-855, https://doi.org/10.1038/s41893-019-0364-5. [44]

Baruch-Mordo, S. et al. (2019), "From Paris to practice: sustainable implementation of renewable energy goals", *Environmental Research Letters*, Vol. 14/2, p. 024013, https://doi.org/10.1088/1748-9326/aaf6e0. [30]

Bennun, L. et al. (2021), *Mitigating biodiversity impacts associated with solar and wind energy development: guidelines for project developers*, IUCN, International Union for Conservation of Nature, https://doi.org/10.2305/iucn.ch.2021.04.en. [26]

Bergmann, A., N. Hanley and R. Wright (2006), "Valuing the attributes of renewable energy investments", *Energy Policy*, Vol. 34/9, pp. 1004-1014, https://doi.org/10.1016/j.enpol.2004.08.035. [94]

Beria, P., I. Maltese and I. Mariotti (2012), "Multicriteria versus Cost Benefit Analysis: a comparative perspective in the assessment of sustainable mobility", *European Transport Research Review*, Vol. 4/3, pp. 137-152, https://doi.org/10.1007/s12544-012-0074-9. [84]

Blaydes, H. et al. (2021), "Opportunities to enhance pollinator biodiversity in solar parks", *Renewable and Sustainable Energy Reviews*, Vol. 145, p. 111065, https://doi.org/10.1016/j.rser.2021.111065. [51]

Blyth-Skyrme, R. (2011), *Benefits and disadvantages of co-locating windfarms and marine conservation zones, with a focus on commercial fishing*, http://www.offshorewind.co.uk (accessed on 29 June 2022). [63]

Bream, H. (2018), *Nevada's Mines Could Hold Key to Question 6 Energy Standard*, Las Vegas Review-Journal, https://www.reviewjournal.com/business/energy/nevadas-mines-could-hold-key-to-question-6-energy-standard1524958/. [36]

Buhl-Mortensen, L. et al. (2017), "Maritime ecosystem-based management in practice: Lessons learned from the application of a generic spatial planning framework in Europe", *Marine Policy*, Vol. 75, pp. 174-186, https://doi.org/10.1016/j.marpol.2016.01.024. [128]

Capellán-Pérez, I., C. de Castro and I. Arto (2017), "Assessing vulnerabilities and limits in the transition to renewable energies: Land requirements under 100% solar energy scenarios", *Renewable and Sustainable Energy Reviews*, Vol. 77, pp. 760-782, https://doi.org/10.1016/j.rser.2017.03.137. [19]

CBD (2004), *CBD Guidelines for the Ecosystem Approach*, https://www.cbd.int/doc/publications/ea-text-en.pdf. [130]

CMS (2022), *Energy Task Force | CMS*, https://www.cms.int/en/taskforce/energy-task-force (accessed on 27 July 2022). [123]

Creutzig, F. et al. (2021), "Considering sustainability thresholds for BECCS in IPCC and biodiversity assessments", *GCB Bioenergy*, Vol. 13/4, pp. 510-515, https://doi.org/10.1111/gcbb.12798. [8]

Creutzig, F. et al. (2021), "Demand-side solutions to climate change mitigation consistent with high levels of well-being", *Nature Climate Change*, Vol. 12/1, pp. 36-46, https://doi.org/10.1038/s41558-021-01219-y. [10]

de Rijk, S. et al. (2021), *Potential synergy between solar energy and biodiversity*, Copernicus GmbH, https://doi.org/10.5194/egusphere-egu21-7040. [55]

Degraer, S. et al. (2020), "Offshore Wind Farm Artificial Reefs Affect Ecosystem Structure and Functioning: A Synthesis", *Oceanography*, Vol. 33/4, pp. 48-57, https://doi.org/10.5670/oceanog.2020.405. [65]

Deshmukh, R. et al. (2019), "Geospatial and techno-economic analysis of wind and solar resources in India", *Renewable Energy*, Vol. 134, pp. 947-960, https://doi.org/10.1016/j.renene.2018.11.073. [38]

Didderen, K., J. Bergsma and P. Kamermans (n.d.), "Offshore flat oyster pilot Luchterduinen wind farm Results campaign 2 (July 2019) and lessons learned Status: draft", *2019*, http://www.buwa.nl (accessed on 18 July 2022). [57]

DOE (2022), *InSPIRE - US Department of Energy*, https://openei.org/wiki/InSPIRE (accessed on 11 July 2022). [61]

Domínguez-Tejo, E. et al. (2016), "Marine Spatial Planning advancing the Ecosystem-Based Approach to coastal zone management: A review", *Marine Policy*, Vol. 72, pp. 115-130, https://doi.org/10.1016/j.marpol.2016.06.023. [127]

Dreves, H. (2019), *Beneath Solar Panels, the Seeds of Opportunity Sprout | News | NREL*, NREL, https://www.nrel.gov/news/features/2019/beneath-solar-panels-the-seeds-of-opportunity-sprout.html (accessed on 11 July 2022). [60]

Drutschinin, A. et al. (2015), "Biodiversity and Development Co-operation", *OECD Development Co-operation Working Papers*, No. 21, OECD Publishing, Paris, https://doi.org/10.1787/5js1sqkvts0v-en. [120]

Dunnett, S. et al. (2022), "Predicted wind and solar energy expansion has minimal overlap with multiple conservation priorities across global regions", *Proceedings of the National Academy of Sciences*, Vol. 119/6, https://doi.org/10.1073/pnas.2104764119. [28]

EC (2022), *Maritime Spatial Plan for the German EEZ in the North Sea | The European Maritime Spatial Planning Platform*, https://maritime-spatial-planning.ec.europa.eu/practices/maritime-spatial-plan-german-eez-north-sea (accessed on 18 July 2022). [77]

ENTSO-E (2019), *3 rd ENTSO-E Guideline for Cost Benefit Analysis of Grid Development Projects Draft version*, https://eepublicdownloads.entsoe.eu/clean-documents/tyndp-documents/Cost%20Benefit%20Analysis/191023_CBA3_Draft%20for%20consultation.pdf (accessed on 13 July 2022). [107]

European Commission (2020), *Commission notice Guidance document on wind energy developments and EU nature legislation Commission notice Guidance document on wind energy developments and EU nature legislation Guidance document on wind energy developments and EU Nature Legislation*, European Commission (C2020) 7730 final, Brussels, https://ec.europa.eu/environment/nature/natura2000/management/docs/wind_farms_en.pdf (accessed on 3 May 2022). [39]

Ferrer, M. et al. (2020), "Transporting Biodiversity Using Transmission Power Lines as Stepping-Stones?", *Diversity*, Vol. 12/11, p. 439, https://doi.org/10.3390/d12110439. [48]

France (2022), *Country response to OECD questionnaire on mainstreaming biodiversity into renewable energy*. [72]

Gauld, J. et al. (2022), "Hotspots in the grid: Avian sensitivity and vulnerability to collision risk from energy infrastructure interactions in Europe and North Africa", *Journal of Applied Ecology*, Vol. 59/6, pp. 1496-1512, https://doi.org/10.1111/1365-2664.14160. [80]

Goldthau, A. (2014), "Rethinking the governance of energy infrastructure: Scale, decentralization and polycentrism", *Energy Research & Social Science*, Vol. 1, pp. 134-140, https://doi.org/10.1016/j.erss.2014.02.009. [113]

Grubler, A. et al. (2018), "A low energy demand scenario for meeting the 1.5 °C target and sustainable development goals without negative emission technologies", *Nature Energy*, Vol. 3/6, pp. 515-527, https://doi.org/10.1038/s41560-018-0172-6. [3]

Halada, L. et al. (2011), "Which habitats of European importance depend on agricultural practices?", *Biodiversity and Conservation*, Vol. 20/11, pp. 2365-2378, https://doi.org/10.1007/s10531-011-9989-z. [33]

Hein, C., J. Whiting and J. Page (2022), *About WREN | Tethys*, https://tethys.pnnl.gov/about-wren (accessed on 27 July 2022). [124]

HM Treasury (2022), *Guidance The Green Book*, https://www.gov.uk/government/publications/the-green-book-appraisal-and-evaluation-in-central-governent/the-green-book-2020#a1-non-market-valuation-and-unmonetisable-values. [87]

Hof, C. et al. (2018), “Bioenergy cropland expansion may offset positive effects of climate change mitigation for global vertebrate diversity”, *Proceedings of the National Academy of Sciences*, Vol. 115/52, pp. 13294-13299, https://doi.org/10.1073/pnas.1807745115. [6]

Höfer, T. et al. (2014), “Wind Farm Siting Using a Spatial Analytic Hierarchy Process Approach: A Case Study of the Sttdteregion Aachen”, *SSRN Electronic Journal*, https://doi.org/10.2139/ssrn.2548520. [105]

Hoffacker, M., M. Allen and R. Hernandez (2017), “Land-Sparing Opportunities for Solar Energy Development in Agricultural Landscapes: A Case Study of the Great Central Valley, CA, United States”, *Environmental Science & Technology*, Vol. 51/24, pp. 14472-14482, https://doi.org/10.1021/acs.est.7b05110. [49]

Holland, R. et al. (2019), “The influence of the global electric power system on terrestrial biodiversity”, *Proceedings of the National Academy of Sciences*, Vol. 116/51, pp. 26078-26084, https://doi.org/10.1073/PNAS.1909269116. [118]

IEA (2021), *Do we need to change our behaviour to reach net zero by 2050? – Analysis - IEA*, https://www.iea.org/articles/do-we-need-to-change-our-behaviour-to-reach-net-zero-by-2050 (accessed on 24 June 2022). [9]

Infrastructure Australia (2021), *Guide to economic appraisal Technical guide of the Assessment Framework*, Australian Government, https://www.infrastructureaustralia.gov.au/sites/default/files/2021-07/Assessment%20Framework%202021%20Guide%20to%20economic%20appraisal.pdf (accessed on 4 August 2022). [86]

IPCC (2022), *Climate Change 2022: Mitigation of Climate Change. Working Group III contribution to the Sixth Assessment Report of the Intergovernmental Panel on Climate Change*. [5]

James, J. and E. Hansen (2017), *Prospects for Large-scale Solar on Degraded Land in West Virginia*, Appalachian Stewardship Foundation and Downstream Strategies, http://www.downstreamstrategies.com/documents/reports_publication/utility-scale-solar-report_2-21-17.pdf (accessed on 4 March 2022). [31]

Kim, C., S. Jang and T. Kim (2018), “Site selection for offshore wind farms in the southwest coast of South Korea”, *Renewable Energy*, Vol. 120, pp. 151-162, https://doi.org/10.1016/j.renene.2017.12.081. [93]

Klein, S. and S. Whalley (2015), “Comparing the sustainability of U.S. electricity options through multi-criteria decision analysis”, *Energy Policy*, Vol. 79, pp. 127-149, https://doi.org/10.1016/j.enpol.2015.01.007. [100]

Koelman, M., T. Hartmann and T. Spit (2021), “When tensions become conflicts: wind turbine policy implementation and development in the Netherlands”, *Journal of Environmental Planning and Management*, Vol. 65/3, pp. 375-397, https://doi.org/10.1080/09640568.2021.1885018. [114]

Larsson, S. and L. Emmelin (2015), “Objectively best or most acceptable? Expert and lay knowledge in Swedish wind power permit processes”, *Journal of Environmental Planning and Management*, Vol. 59/8, pp. 1360-1376, https://doi.org/10.1080/09640568.2015.1076383. [117]

Lee, N., F. Flores-Espino and D. Hurlbut (2017), *Renewable Energy Zone (REZ) Transmission Planning Process: A Guidebook for Practitioners*, USAID; NREL. [73]

Luderer, G. et al. (2019), "Environmental co-benefits and adverse side-effects of alternative power sector decarbonization strategies", *Nature Communications 2019 10:1*, Vol. 10/1, pp. 1-13, https://doi.org/10.1038/s41467-019-13067-8. [13]

Macgregor, C. et al. (2022), "Brownfield sites promote biodiversity at a landscape scale", *Science of The Total Environment*, Vol. 804, p. 150162, https://doi.org/10.1016/j.scitotenv.2021.150162. [34]

Macknick, J., B. Beatty and G. Hill (2013), "Overview of Opportunities for Co-Location of Solar Energy Technologies and Vegetation", http://www.nrel.gov/publications. (accessed on 29 June 2022). [64]

Magar, V. (ed.) (2016), "Reconciling Biodiversity Conservation and Widespread Deployment of Renewable Energy Technologies in the UK", *PLOS ONE*, Vol. 11/5, p. e0150956, https://doi.org/10.1371/journal.pone.0150956. [21]

Mai, T. et al. (2021), "Interactions of wind energy project siting, wind resource potential, and the evolution of the U.S. power system", *Energy*, Vol. 223, p. 119998, https://doi.org/10.1016/j.energy.2021.119998. [18]

McKenney, B. (2020), *Clean and Green Pathways for the Global Renewable Energy Buildout*. [71]

McManamay, R., C. Vernon and H. Jager (2021), "Global Biodiversity Implications of Alternative Electrification Strategies Under the Shared Socioeconomic Pathways", *Biological Conservation*, Vol. 260, p. 109234, https://doi.org/10.1016/J.BIOCON.2021.109234. [24]

Michalena, E. and J. Hills (2012), "Renewable energy issues and implementation of European energy policy: The missing generation?", *Energy Policy*, Vol. 45, pp. 201-216, https://doi.org/10.1016/j.enpol.2012.02.021. [115]

Moran, E. et al. (2018), "Sustainable hydropower in the 21st century", *Proceedings of the National Academy of Sciences*, Vol. 115/47, pp. 11891-11898, https://doi.org/10.1073/pnas.1809426115. [23]

Newbold, T. et al. (2015), "Global effects of land use on local terrestrial biodiversity", *Nature*, Vol. 520/7545, pp. 45-50, https://doi.org/10.1038/nature14324. [29]

Nilsson, C. et al. (2005), "Fragmentation and Flow Regulation of the World's Large River Systems", *Science*, Vol. 308/5720, pp. 405-408, https://doi.org/10.1126/science.1107887. [15]

Nock, D. and E. Baker (2019), "Holistic multi-criteria decision analysis evaluation of sustainable electric generation portfolios: New England case study", *Applied Energy*, Vol. 242, pp. 655-673, https://doi.org/10.1016/j.apenergy.2019.03.019. [99]

Norway Directorate for Financial Management (2018), *Veileder i samfunnsøkonomiske*, https://dfo.no/sites/default/files/fagomr%C3%A5der/Utredninger/Veileder-i-samfunnsokonomiske-analyser.pdf#page=24&zoom=100,0,0. [88]

NVE (2022), *Knowledge base on the effects of wind power on land*, https://www.nve.no/energi/energisystem/vindkraft/kunnskapsgrunnlag-om-virkninger-av-vindkraft-paa-land/. [110]

OECD (2023), *A Decade of Development Finance for Biodiversity*, OECD Publishing, Paris, https://doi.org/10.1787/e6c182aa-en. [122]

OECD (2023), *OECD Practical Tool on Environmental Due Diligence In Mineral Supply Chains*, https://www.umweltbundesamt.de/sites/default/files/medien/6232/dokumente/211102_factsheet_practicaltoolenvddmineralsc.pdf. [119]

OECD (2021), *Transport Strategies for Net-Zero Systems by Design*, OECD Publishing, Paris, https://doi.org/10.1787/0a20f779-en. [12]

OECD (2020), *Towards Sustainable Land Use: Aligning Biodiversity, Climate and Food Policies*, OECD Publishing, Paris, https://doi.org/10.1787/3809b6a1-en. [112]

OECD (2019), *Accelerating Climate Action: Refocusing Policies through a Well-being Lens*, OECD Publishing, Paris, https://doi.org/10.1787/2f4c8c9a-en. [11]

OECD (2019), "Green triangular co-operation: An accelerator to sustainable development", *OECD Development Policy Papers*, No. 21, OECD Publishing, Paris, https://doi.org/10.1787/d81d884a-en. [121]

OECD (2018), *Cost-Benefit Analysis and the Environment: Further Developments and Policy Use*, OECD Publishing, Paris, https://doi.org/10.1787/9789264085169-en. [85]

OECD (2018), *Mainstreaming Biodiversity for Sustainable Development*, OECD Publishing, Paris, https://doi.org/10.1787/9789264303201-en. [111]

OECD (2017), *Marine Spatial Planning Assessing net benefits and improving effectiveness*, https://www.oecd.org/greengrowth/GGSD_2017_Issue%20Paper_Marine%20Spatial%20Planning.pdf (accessed on 25 July 2022). [76]

OECD (2016), *Biodiversity Offsets: Effective Design and Implementation*, OECD Publishing, Paris, https://doi.org/10.1787/9789264222519-en. [25]

Opperman, J. et al. (2019), *Connected and Flowing: A renewable future for rivers, climate and people*, WWF and TNC, Washington D.C. [22]

Pascaris, A. et al. (2022), "Do agrivoltaics improve public support for solar? A survey on perceptions, preferences, and priorities", *Green Technology, Resilience, and Sustainability*, Vol. 2/1, https://doi.org/10.1007/s44173-022-00007-x. [46]

Passoni, G. et al. (2017), "Framework for strategic wind farm site prioritisation based on modelled wolf reproduction habitat in Croatia", *European Journal of Wildlife Research*, Vol. 63/2, https://doi.org/10.1007/s10344-017-1092-7. [81]

Pfeifenberger et al. (2023), *The Benefit and Urgency of Planned Offshore Transmission: Reducing the Costs of and Barriers to Achieving U.S. Clean Energy Goals*, https://www.brattle.com/wp-content/uploads/2023/01/Brattle-OSW-Transmission-Report_Jan-24-2023.pdf (accessed on 28 February 2023). [108]

Pitts-Singer, T. (ed.) (2021), "Can Solar Energy Fuel Pollinator Conservation?", *Environmental Entomology*, Vol. 50/4, pp. 757-761, https://doi.org/10.1093/ee/nvab041. [54]

Pojadas, D. and M. Abundo (2022), "A spatial cost-benefit-externality modelling framework for siting of variable renewable energy farms: A case in Bohol, Philippines", *Renewable Energy*, Vol. 181, pp. 1177-1187, https://doi.org/10.1016/j.renene.2021.09.114. [90]

Pörtner et al. (2021), *Scientific outcome of the IPBES-IPCC co-sponsored workshop on biodiversity and climate change*, IPBES Secretariat, https://doi.org/10.5281/zeonodo.4659158. [16]

Pringle, A., R. Handler and J. Pearce (2017), "Aquavoltaics: Synergies for dual use of water area for solar photovoltaic electricity generation and aquaculture", *Renewable and Sustainable Energy Reviews*, Vol. 80, pp. 572-584, https://doi.org/10.1016/j.rser.2017.05.191. [47]

Prusina, I. et al. (2020), *Nature-Inclusive Design: a catalogue for offshore wind infrastructure The Ministry of Agriculture, Nature and Food Quality Final version*, http://www.witteveenbos.com (accessed on 18 July 2022). [58]

Randle-Boggis, R. et al. (2020), "Realising co-benefits for natural capital and ecosystem services from solar parks: A co-developed, evidence-based approach", *Renewable and Sustainable Energy Reviews*, Vol. 125, p. 109775, https://doi.org/10.1016/j.rser.2020.109775. [50]

Ravi, S. et al. (2016), "Colocation opportunities for large solar infrastructures and agriculture in drylands", *Applied Energy*, Vol. 165, pp. 383-392, https://doi.org/10.1016/j.apenergy.2015.12.078. [62]

RCREEE (2015), *Active Turbine Management Program*, https://www.rcreee.org/project/active-turbine-management-program-atmp/4300 (accessed on 27 June 2022). [109]

Rhodes, J. et al. (2017), "A geographically resolved method to estimate levelized power plant costs with environmental externalities", *Energy Policy*, Vol. 102, pp. 491-499, https://doi.org/10.1016/j.enpol.2016.12.025. [89]

Rijkswaterstaat (2020), *LIFE13ENV/NL/000971 FINAL Report Solar Highways Project Data Beneficiary Data*, http://solarhighways.eu (accessed on 18 April 2023). [40]

Rio Tinto (2012), *Rio Tinto and biodiversity: Working towards net positive impa.* [27]

Rouhani, O. et al. (2016), "Cost-benefit analysis of various California renewable portfolio standard targets: Is a 33% RPS optimal?", *Renewable and Sustainable Energy Reviews*, Vol. 62, pp. 1122-1132, https://doi.org/10.1016/j.rser.2016.05.049. [91]

RTE (2022), *Futurs énergétiques 2050 : les scénarios de mix de production à l'étude permettant d'atteindre la neutralité carbone à l'horizon 2050. Chapitre 12: L'analyse environnementale.* [2]

Saarikoski, H. et al. (2016), "Multi-Criteria Decision Analysis and Cost-Benefit Analysis: Comparing alternative frameworks for integrated valuation of ecosystem services", *Ecosystem Services*, Vol. 22, pp. 238-249, https://doi.org/10.1016/j.ecoser.2016.10.014. [98]

Saikia, L. (ed.) (2022), "Land-use intensity of electricity production and tomorrow's energy landscape", *PLOS ONE*, Vol. 17/7, p. e0270155, https://doi.org/10.1371/journal.pone.0270155. [14]

Sánchez-Lozano, J., M. García-Cascales and M. Lamata (2016), "Comparative TOPSIS-ELECTRE TRI methods for optimal sites for photovoltaic solar farms. Case study in Spain", *Journal of Cleaner Production*, Vol. 127, pp. 387-398, https://doi.org/10.1016/j.jclepro.2016.04.005. [102]

Santangeli, A. et al. (2015), "Global change synergies and trade-offs between renewable energy and biodiversity", *GCB Bioenergy*, Vol. 8/5, pp. 941-951, https://doi.org/10.1111/gcbb.12299. [20]

Schupp, M. et al. (2021), "Fishing within offshore wind farms in the North Sea: Stakeholder perspectives for multi-use from Scotland and Germany", *Journal of Environmental Management*, Vol. 279, p. 111762, https://doi.org/10.1016/j.jenvman.2020.111762. [67]

Semeraro, T. et al. (2018), "Planning ground based utility scale solar energy as green infrastructure to enhance ecosystem services", *Energy Policy*, Vol. 117, pp. 218-227, https://doi.org/10.1016/j.enpol.2018.01.050. [52]

Shorabeh, S. et al. (2019), "A risk-based multi-criteria spatial decision analysis for solar power plant site selection in different climates: A case study in Iran", *Renewable Energy*, Vol. 143, pp. 958-973, https://doi.org/10.1016/j.renene.2019.05.063. [101]

Siegner, K. et al. (2019), *Maximizing land-use benefits from utility scale solar: A cost-benefit analysis of pollinator-friendly solar in Minnesota*, Yale Center for Business and the Environment, https://cbey.yale.edu/sites/default/files/2019-12/MaximizingLandUseBenefitsFromUtility-ScaleSolar_0.pdf (accessed on 6 October 2021). [97]

Sinha, P. et al. (2018), "Best Practices in Responsible Land Use for Improving Biodiversity at a Utility-Scale Solar Facility", *Case Studies in the Environment*, Vol. 2/1, pp. 1-12, https://doi.org/10.1525/cse.2018.001123. [53]

Smith, P. et al. (2019), "Land-Management Options for Greenhouse Gas Removal and Their Impacts on Ecosystem Services and the Sustainable Development Goals", *Annual Review of Environment and Resources*, Vol. 44/1, pp. 255-286, https://doi.org/10.1146/annurev-environ-101718-033129. [7]

Spain (2020), *Environmental Zoning for Implementing Renewable Energy: Wind and Solar*, https://www.miteco.gob.es/es/calidad-y-evaluacion-ambiental/temas/evaluacion-ambiental/zonificacion_ambiental_energias_renovables.aspx (accessed on 17 April 2023). [82]

Steins, N. et al. (2021), "Combining offshore wind farms, nature conservation and seafood: Lessons from a Dutch community of practice", *Marine Policy*, Vol. 126, p. 104371, https://doi.org/10.1016/j.marpol.2020.104371. [68]

Stelzenmüller, V. et al. (2016), "Co-location of passive gear fisheries in offshore wind farms in the German EEZ of the North Sea: A first socio-economic scoping", *Journal of Environmental Management*, Vol. 183, pp. 794-805, https://doi.org/10.1016/j.jenvman.2016.08.027. [66]

Sun, Y. et al. (2013), "GIS-based approach for potential analysis of solar PV generation at the regional scale: A case study of Fujian Province", *Energy Policy*, Vol. 58, pp. 248-259, https://doi.org/10.1016/j.enpol.2013.03.002. [92]

TNC (2022), *Site Renewables Right: Accelerating a Clean and Green Renewable Energy Buildout in the Central United States*, https://www.nature.org/content/dam/tnc/nature/en/documents/SRR_Methods_20220202_LR.pdf. [83]

TNC (2020), *Mining the Sun: How Nevada and West Virginia are Reclaiming Former Mine Lands with Solar Panels*, https://www.nature.org/en-us/what-we-do/our-priorities/tackle-climate-change/climate-change-stories/nevada-west-virginia-solar-energy-former-mines/. [32]

Tran, T. and M. Egermann (2022), "Land-use implications of energy transition pathways towards decarbonisation – Comparing the footprints of Vietnam, New Zealand and Finland", *Energy Policy*, Vol. 166, p. 112951, https://doi.org/10.1016/j.enpol.2022.112951. [4]

Tucker, G., F. Quétier and W. Wende (2020), *uidance on achieving no net loss or net gain of biodiversity and ecosystem services. Report to the European Commission, DG Environment on Contract ENV.B.2/SER/2016/0018*, Institute for European Environmental Policy, Brussels. [69]

Turner, R. (2016), "The 'Balance Sheet' Approach Within Adaptive Management for Ecosystem Services", in *Routledge Handbook of Ecosystem Services*, Routledge, https://doi.org/10.4324/9781315775302-25. [106]

Turner, R. (2007), "Limits to CBA in UK and European environmental policy: retrospects and future prospects", *Environmental and Resource Economics*, Vol. 37/1, pp. 253-269, https://doi.org/10.1007/s10640-007-9119-2. [96]

U.S. (2022), *United States Response to OECD Working Party on Biodiversity Water and Ecosystems Questionnaire*. [129]

US Bureau of Land Management (2016), "Executive Summary for the Record of Decision for the Land Use Plan Amendment to the California Desert Conservation Plan, Bishop Resource Management Plan, and Bakersfield Resource Management Plan", https://eplanning.blm.gov/public_projects/lup/66459/133459/163123/DRECP_BLM_ROD_Executive_Summary.pdf (accessed on 3 March 2022). [75]

US Bureau of Land Management (2015), *Desert Renewable Energy Conservation Plan: Proposed Land Use Plan Amendment and Final Environmental Impacts Statement*. [74]

US DOI (2015), *Interior Department Approves First Solar Energy Zone Projects*, https://www.doi.gov/pressreleases/interior-department-approves-first-solar-energy-zone-projects (accessed on 15 July 2022). [70]

van de Ven, D. et al. (2021), "The potential land requirements and related land use change emissions of solar energy", *Scientific Reports*, Vol. 11/1, https://doi.org/10.1038/s41598-021-82042-5. [41]

van Ouwerkerk, J. et al. (2022), "Impacts of power sector model features on optimal capacity expansion: A comparative study", *Renewable and Sustainable Energy Reviews*, Vol. 157, p. 112004, https://doi.org/10.1016/j.rser.2021.112004. [17]

Villarini, M. (ed.) (2018), "Remarkable agrivoltaic influence on soil moisture, micrometeorology and water-use efficiency", *PLOS ONE*, Vol. 13/11, p. e0203256, https://doi.org/10.1371/journal.pone.0203256. [42]

Vrees, L. (2021), "Adaptive marine spatial planning in the Netherlands sector of the North Sea", *Marine Policy*, Vol. 132, p. 103418, https://doi.org/10.1016/j.marpol.2019.01.007. [78]

Waardenburg, B. (2020), *Biodiversity enhancement in offshore wind farms Options for biodiversity enhancement in offshore wind farms Knowledge base for the implementation of the Rich North Sea Programme*, https://www.buwa.nl/fileadmin/buwa_upload/Bureau_Waardenburg_rapporten/2020/18-0660_The_Rich_North_Sea-_options_for_biodiversity_enhancement_in_OWFs_07022020-reduced.pdf (accessed on 18 July 2022). [59]

WaterProof Marine Consultancy and Services (2020), *Artificial reef structures North Sea: Structure stability analyses in Borssele OWF and NSIL*, https://de-rijke-noordzee-production.s3-eu-west-1.amazonaws.com/Website/WP2020_1218_R1r2_Stability_eco_structures_Northsea_final.pdf (accessed on 18 July 2022). [56]

Watson, J. and M. Hudson (2015), "Regional Scale wind farm and solar farm suitability assessment using GIS-assisted multi-criteria evaluation", *Landscape and Urban Planning*, Vol. 138, pp. 20-31, https://doi.org/10.1016/j.landurbplan.2015.02.001. [104]

Winemiller, K. et al. (2016), "Balancing hydropower and biodiversity in the Amazon, Congo, and Mekong", *Science*, Vol. 351/6269, pp. 128-129, https://doi.org/10.1126/science.aac7082. [126]

Wu, G. et al. (2019), "Power of Place Land Conservation and Clean Energy Pathways for California", https://www.scienceforconservation.org/assets/downloads/Technical_Report_Power_of_Place.pdf (accessed on 30 March 2022). [1]

Zahid, F. et al. (2021), "Wind farms selection using geospatial technologies and energy generation capacity in Gwadar", *Energy Reports*, Vol. 7, pp. 5857-5870, https://doi.org/10.1016/j.egyr.2021.08.165. [103]

Zarfl, C. et al. (2019), "Future large hydropower dams impact global freshwater megafauna", *Scientific Reports*, Vol. 9/1, https://doi.org/10.1038/s41598-019-54980-8. [125]

Zoellner, J., P. Schweizer-Ries and C. Wemheuer (2008), "Public acceptance of renewable energies: Results from case studies in Germany", *Energy Policy*, Vol. 36/11, pp. 4136-4141, https://doi.org/10.1016/j.enpol.2008.06.026. [116]

Notes

[1] Lifestyles change to increase energy sufficiency in terms of end-uses and consumption (less individual travel, favouring soft mobility and mass transport, less consumption of manufactured goods, sharing economy, lower set point temperatures for heating, increase in remote working, digital sustainability, etc.), resulting in an overall reduction in energy needs, and thus electricity needs.

[2] Gradual electrification (substitution for fossil fuels) and ambitious targets for energy efficiency (NLCS assumption). Assumes continued economic growth (+1.3% per year from 2030) and demographic growth (INSEE's low fertility scenario). The baseline trajectory assumes a high degree of efficacy of public policies and plans (stimulus, hydrogen, industry). The manufacturing industry expands, and its share of GDP ceases to decrease. Building renovation is factored in but so is the related rebound effect.

[3] The predecessor of the MacKay Carbon Calculator.

[4] Not only does hydropower tend to be more harmful than solar and wind, much of the untapped potential for hydropower is found in river basins with exceptionally high biodiversity (Zarfl et al., 2019[125]; Winemiller et al., 2016[126]).

[5] Upstream impacts from mining for minerals used in the component parts may still occur.

[6] Increasing turbine and hub height has facilitated this (see Chapter 3).

[7] The ecosystem approach is an interdisciplinary management approach that recognises the complexity of ecological systems and integrates social, ecological, and governance principles to achieve equitable and sustainable natural resource use (Domínguez-Tejo et al., 2016[127]; Buhl-Mortensen et al., 2017[128])· For guidelines on the Ecosystem Approach see CBD Secretariat (2004[130]), CBD Guidelines for the Ecosystem Approach, Montreal.

[8] Not all of these approaches have been operationalised, some were developed as academic demonstrations.

[9] CBA for renewable energy is not standard practice globally. It can also be challenging owing to the significant and rapid fluctuations and uncertainties in energy prices.

[10] Wildlife designations included: UK Sites of Special Scientific Interest, National and Local Nature Reserves; European Special Protection Areas and Special Areas of Conservation; and international Ramsar site and "landscape designations", comprising National Parks and Areas of Outstanding Natural Beauty. Other constraint areas included, historically important sites, residential area, agricultural land classification and aspect (direction of slope) and slope (gradient of land).

[11] The guideline was developed in compliance with the requirements of the EU Regulation (EU) 347/2013, which aims to ensure a common framework for multi-criteria cost-benefit analysis (CBA) for ENTSO-E Ten Year Network Development Plan (TYNDP) project.

[12] Sensitive areas are considered to be land protected under the following Directives or International Laws: Habitats Directive (92/43/EEC); Birds Directive (2009/147/EC); RAMSAR site; IUCN key biodiversity areas; Marine Strategy Framework Directive (2008/56/EC); Other nature protection areas under national law.

[13] In the US, for example, regulations differ from one state to another, affecting the location and practices of renewable power projects (U.S., 2022[129]). Most projects are in states and on lands that have fewer regulatory requirements for renewable power companies to build and operate.

5 Policy instruments for reconciling biodiversity protection and renewable power expansion

This chapter presents policy instruments for integrating biodiversity considerations into the deployment of renewable power infrastructure. Drawing on examples, it explores how regulatory, economic, voluntary and information instruments are being used to address biodiversity impacts from solar, wind and power line developments. The chapter also highlights key design elements to promote effective application of these instruments.

Governments use various policy instruments to manage the impacts of renewable energy on biodiversity. A carefully designed policy mix can ensure that renewable power projects effectively mitigate adverse impacts on biodiversity and seek positive outcomes for nature. An effective mix will likely require a combination of regulatory (command-and-control), economic and other instruments (e.g., voluntary approaches and information instruments) (see Table 5.1 for an overview). What constitutes an appropriate mix will differ across countries and jurisdictions.

This chapter provides an overview of the policy instruments used to address biodiversity impacts from renewable energy deployment. It provides examples and highlights the strengths, challenges and limitations of the various instruments. The analysis focuses on instruments for managing the impacts at the construction and operation stage of the renewable energy life cycle. Some of the instruments covered here (e.g., environmental impact assessments, biodiversity offsets, responsible business conduct and environmental labelling schemes) can also be tailored to address the negative biodiversity impacts associated with sourcing materials for renewable energy infrastructure and eventual decommissioning of infrastructure. However, additional policy instruments are required to address the full life cycle impacts of renewable power development on biodiversity and ensure a biodiversity-aligned transition to low-emissions electricity systems (Box 5.1).

Table 5.1. Policy options for aligning renewable power and biodiversity objectives

Regulatory (command-and-control)	Economic instruments	Information instruments and voluntary approaches
Spatial planning and renewable energy zoning	Biodiversity offsets	Procurement policies, tender processes and power purchase agreements that integrate biodiversity criteria
Environmental licensing and permitting requirements	Subsidies (e.g. for RDD or renewable projects seeking positive biodiversity outcomes)	Voluntary corporate commitments (e.g. no net loss) and investor performance standards
Strategic environmental assessment (SEA)	Payments for ecosystem services	Voluntary industry guidelines
Environmental impact assessment (EIA)		Ecolabels and certification (e.g. biodiversity-friendly solar energy certification)
Renewable energy design and operation standards		
Monitoring, data sharing and disclosure requirements		
Responsible business conduct (RBC) due diligence requirements		

Note: Some instruments categorised here as regulatory may also be voluntary in some countries (e.g., RBC due diligence requirements).
Source: Author.

Box 5.1. Resource efficiency and circularity for a biodiversity-aligned transition to low-emissions electricity

The growing demand for materials to construct renewable energy, power lines and associated low-emission technologies (e.g., batteries and electric vehicles), risks adding significant pressure to biodiversity owing to increased mining pressure and waste from decommissioned infrastructure. As outlined in Chapter 3, mineral demand (tonnes) for low-emission technologies increases six-fold from 2020-40 in IEA's net-zero scenario 2050. Cumulative waste from solar PV alone is projected to reach 60-78 million tonnes by 2050.

Governments can help reduce mineral resource demand and environmentally harmful waste by promoting resource efficiency and material circularity. This requires a life-cycle approach with policies that promote resource productivity, material recovery, sustainable materials management and the 3Rs – reduce, reuse, recycle. Examples of policy options include:

- Extended producer responsibility (EPR): EPR is an environmental policy approach in which a producer's responsibility for a product is extended to the post-consumer stage of the product's lifecycle. The approach seeks to increase recovery rates and decrease waste generation and leakage by transferring end-of-life management costs from the general public to producers and consumers of targeted products. The EU Waste Electrical and Electronic Equipment Directive, for example, requires all producers supplying PV panels to the EU market to finance the costs of collecting and recycling end-of-life PV panels put on the market in Europe.
- Taxes on virgin material and landfill: Taxes on virgin materials (or materials that are toxic or difficult to recycle) incentivise efficient resource use by increasing the cost of extracting and using primary natural resources and raw materials. Resource efficiency may be realised through eco-designs that are less material-intensive, use less toxic materials and/or use secondary (recycled) materials. Landfill taxes increase the cost of waste disposal, thereby helping divert waste flows from landfills towards reuse or recycling.
- Regulatory measures: Regulatory measures such as minimum recycled content, product standards, lifetime warranties, bans and restrictions can promote resource efficiency and circularity for the component parts of renewable energy infrastructure and other low-emissions technologies.

Source: (IRENA and IEA-PVPS, 2016[1]), End-Of-Life Management: Solar Photovoltaic Panels; (OECD, 2022[2]), Towards a More Resource-Efficient and Circular Economy: The Role of G20 OECD-G20-Towards-a-more-Resource-Efficient-and-Circular-Economy.pdf; (OECD, 2020[3]), Improving Resource Efficiency and the Circularity of Economies for a Greener World, https://doi.org/10.1787/1b38a38f-en; (OECD, 2016[4]), Extended Producer Responsibility: Updated Guidance for Efficient Waste Management, OECD Publishing, Paris https://doi.org/10.1787/9789264256385-en.

5.1. Regulatory instruments

Regulatory (command-and-control) instruments are traditionally the cornerstone of biodiversity policy. They can reduce pressure on biodiversity by dictating which activities are permissible, when, where and how. Regulatory approaches are particularly important where markets cannot provide price signals to organisations to reflect the costs of polluting behaviour, or where strict control is required to safeguard biodiversity and avoid ecosystem collapse or species extinctions (e.g., protected areas or outright bans on development in sites of particularly high ecological sensitivity or value).

Generally, regulatory instruments should be i) closely targeted to the policy goal; ii) stringent enough for the benefits to outweigh the cost; iii) stable enough to give investors confidence; iv) sufficiently flexible to foster novel solutions; and v) updated regularly to provide incentives for continuous innovation and to reflect the latest science and knowledge (OECD, 2012[5]).

This section examines several regulatory instruments applied to renewable energy developments: environmental permitting, SEA and EIA, requirements for monitoring, data-sharing and disclosure, due diligence requirements, and infrastructure design and operations standards. Spatial planning was discussed in Chapter 4 and is therefore not covered here.

5.1.1. Environmental permits

In most countries, developers must obtain permits to construct and operate utility scale renewable power infrastructure, however, the process, requirements and timelines vary considerably across jurisdictions. Often developers must seek consent to develop or operate facilities from multiple government agencies, at national and subnational levels. As many as ten permits may be required for constructing and operating a solar power facility and as many as twenty for an offshore wind farm, depending on the project's location, technology, scale and associated environmental risk (Energy Transitions Commission, 2023[6]). For example, in the US, a project that may impact species protected under the Endangered Species Act or the Bald and Gold Eagle Protection Act has an additional requirement to obtain an incidental take permit as part of the permitting process (Box 5.2) (Sud and Patnaik, 2022[7]). Some (typically smaller scale) projects may be exempt from permitting requirements as is the case for rooftop solar in Ireland and solar power plants of less than 1 kilovolt (kV) in Norway that can be connected to established low-voltage installations and wind power plants under 1 megawatt (NVE, 2023[8]; Government of Ireland, 2022[9]).

Box 5.2. US Incidental take permits for eagles and collision risk model for estimating fatalities from wind energy projects

Wind developments that risk killing bald or golden eagles in the US may require an Eagle Incidental Take Permit under the Bald and Golden Eagle Protection Act. The permit authorises the "take" of eagles where the take is compatible with the preservation of bald and golden eagles and cannot be practicably avoided. This type of take is considered "incidental take." The U.S. Fish and Wildlife Service (FWS) encourages the development of an Eagle Conservation Plan for wind energy developers when incidental eagle take may occur and provides guidance on developing such a plan – including what information is needed for permitting. As part of the Eagle Conservation Plan Guidance FWS developed a collision risk model to predict the number of golden and bald eagles that may be killed at wind facilities. The collision risk model incorporates existing knowledge of eagle use around a proposed wind facility (exposure) and the probability of an eagle colliding with an operating turbine. This information is meant to reflect how exposure and collision probability can vary across the nation and takes the form of prior probability distributions. These distributions of prior knowledge are combined with project-specific information to predict the number of fatalities expected for a particular wind facility for consideration when issuing eagle incidental take permits to wind-energy facilities. The aim of the collision risk model is to ensure conservation of bald eagles and golden eagles when issuing permits.

Source: (US FWS, 2021[10]), Updated Collision Risk Model for Estimating Eagle Fatalities at Wind Energy Facilities; (FWS, n.d.[11]), Eagle Management, https://www.fws.gov/program/eagle-management/eagle-permits.

By determining whether a project can go ahead and under what conditions, well-designed permitting can help ensure infrastructure is constructed and operated in harmony with nature. It is good practice to issue permits only for projects that are unlikely to have significant adverse impacts on biodiversity. For example, wind farm developers may be required to demonstrate how, through implementation of the mitigation hierarchy, projected bird collision mortalities will be under appropriate mortality thresholds to obtain a permit (Box 5.3). To help ensure permitted projects do not pose a significant risk to biodiversity, governments can require projects to submit environmental risk screening and impact assessments (see EIA). Permitting may be conditional on the adoption of impact mitigation measures such as increasing the cut-in speed to reduce collision risk and a plan for offsetting residual adverse impacts. In the UK, for example, the 2021 Environment Act requires new developments in England seeking a planning permit to demonstrate a 10% increase in biodiversity at or near the project site, measured using Defra's Biodiversity Metric (UK, 2021[12]). Additionally, governments can require an environmental monitoring plan to be submitted before a permit is issued to ensure the project remains compliant with regulations and respects recommendations of an EIA, as well as to identify and address any unexpected impacts (see section 5.1.4).

Box 5.3. Mortality thresholds: Determining acceptable mortality limits for birds and bats from wind farms

Scientific-based information on the potential impact of wind turbine collisions on birds and bats is fundamental for determining whether, and under what conditions, a wind energy project should go ahead. This requires not only an estimate of bird and bat mortality rates, but also how these additional individual mortalities impact the population. Key factors determining the impact severity of additional mortalities include: 1) the size of the impacted population; 2) its current demographic trend; and 3) the species vital rates (survivals and fecundities).

Mortality thresholds are used for wind power projects in several countries and by some financial institutions to determine an acceptable rate of mortality. Two main approaches are used. The first assumes a specified percentage of overall annual natural mortality in the relevant biogeographical population is negligible. In Netherlands, for example, this is 1% for all species while in Belgium it is 1% except for abundant species with favourable conservation status for which the threshold is 5%. In Germany, the "significance" threshold is 0.5%-5%. The second approach is the potential biological removal (PBR) method, which was originally developed for the management of marine mammals but has since been applied to other taxa. The US Marine Mammal Protection Act defines PBR as "the maximum number of animals, not including natural mortalities, that may be removed from a marine mammal stock while allowing that stock to reach or maintain its optimal sustainable population."

Modelling for seven species suggests that commonly used mortality limits may lead to severe population-level impacts. A 1% additional mortality in post-fledging cohorts resulted in a 2%-24% decrease in population after 10 years. An additional mortality of 5% resulted in a 9%-77% decrease after 10 years. Alternative approaches for estimating appropriate mortality limits may be required, such as the one proposed by Schippers et al. 2020.

Source: (Chambert and Besnard, 2021[13]), Assessing the demographic impact of bird collisions with wind turbines. State of the art and methodological recommendations; (Schippers et al., 2020[14]), Mortality limits used in wind energy impact assessment underestimate impacts of wind farms on bird populations.

It is good practice to integrate environmental risk screening and EIA into the permitting process and for permit decisions to reflect the findings and recommendations of the assessment. Conducting an EIA after key decisions have been made and the development permit issued – or ignoring the recommendations of an EIA – limits the effectiveness of an EIA in safeguarding biodiversity. The extent to which the EIA and permitting processes are integrated differs substantially across countries. In some countries, such as Norway, the EIA process for renewable energy permitting is integrated early in the permitting process and is a precursor for submitting a permit request (NVE, 2022[15]; OECD, 2022[16]). In other countries the EIA and permitting processes are not directly linked and authorisation can be given for projects even where the EIA conclusion was negative (OECD, 2021[17]; OECD, 2021[18]).

The criteria determining which projects must conduct environmental risk screening and assessments also varies across countries. The criteria may depend on the type of project, its location or size. For example, in Japan an EIA is required for solar projects of at least 40 MW and wind of at least 10 MW. EIA is applied to solar power projects of 30-40 MW and wind power projects of 7.5-10 MW if deemed necessary after screening. In EU Member States, a renewable energy project in a Natura 2000 site must undergo a pre-assessment (screening) as part of the permitting process. If the screening determines that the project is likely to have a significant effect on biodiversity (Box 5.4), then an Appropriate Assessment is also required (European Commission, 2020[19]). In Indonesia, an EIA is required for transmission lines with capacity of 150 kV or more and solar and wind projects of 10 MW or more (Indonesia, 2012[20]). Careful design of criteria is fundamental to prevent harmful projects being permitted. Such criteria should be clear, consistent with latest science and knowledge, and revisited over time to adjust as necessary.

Box 5.4. Determining whether impacts are "significant" – EU Natura 2000 sites and renewable energy projects

The notion of what is 'significant' needs to be interpreted objectively. The significance of effects should be determined in relation to the specific features and environmental conditions of the protected site affected by the plan or project, taking particular account of the site's conservation objectives and ecological characteristics.

An assessment of significant effects must be based on good science (including best available methods and knowledge) and reliable data, be precautionary and, if appropriate, it should take into account the opinion of stakeholders, such as NGOs, nature conservation agencies or researchers. The assessment must apply the principle of proportionality, be compatible with the precautionary principle and consider:

- the nature, size and complexity of the plan or project;
- the expected effects, and
- the vulnerability and irreplaceability of the affected EU-protected habitats and species.

Source: Extract from (European Commission, 2020[19]), Commission notice Guidance document on wind energy developments and EU nature legislation Commission notice Guidance document on wind energy developments and EU nature legislation Guidance document on wind energy developments and EU Nature Legislation. https://ec.europa.eu/environment/nature/natura2000/management/docs/wind_farms_en.pdf.

Permitting is recognised universally as a bottleneck for swiftly scaling up renewable power projects (IEA, 2022[21]; IEA, 2022[22]). In some EU Member States, for example, granting permits for large projects can take nine years (EC, 2022[23]). This is concerning given the urgency of phasing out fossil fuels and ensuring energy security. Jurisdictions across the world are therefore revising their permitting processes with an aim to accelerate them (e.g. the REPowerEU Plan (European Commission, 2022[24]); and the Biden-Harris Administration's Permitting Action Plan in the US (White House, 2022[25])).

Accelerating permitting of renewable power projects is important, but it must not increase the risk of significant adverse impacts on biodiversity. Multiple regulatory, administrative and societal barriers lead to slow and costly permitting processes, such as opaque or conflicting rules and regulations, over-complex and bureaucratic processes, lack of transparency, under-resourced permitting authorities and resistance from local communities and other stakeholders (Jack, 2022[26]; Energy Transitions Commission, 2023[6]; EEB, 2022[27]; Ulibarri, Cain and Ajami, 2017[28]). Many solutions to overcome these barriers can be implemented without posing additional risk to biodiversity, for example, establishing a one-stop-shop approach[1], enhancing the capacity of authorities to efficiently process permit requests, setting permitting targets (timelines), developing guidelines to provide clarity to developers on the permitting process and enhancing transparency (and reducing paperwork) through digitisation (Vasconcelos et al., 2022[29]; Energy Transitions Commission, 2023[30]; EC, 2022[31]; Barr et al., 2021[32]).

Moreover, opportunities exist in many countries to reduce delays in renewable energy deployment by smarter and more systematic integration of biodiversity into planning and permitting processes. Key steps governments can take include:

- Ensuring early and ongoing engagement of local communities, environmental experts and other stakeholders: involving stakeholders throughout the project development process can help build awareness and trust, and inform project design, thereby reducing the risk of opposition and associated permitting delays and legal cases (Energy Transitions Commission, 2023[30]; IEA, 2018[33]; Susskind et al., 2022[34]; Pollard and Bennun, 2016[35]).
- Developing spatial plans such as renewable energy zones that explicitly account for biodiversity, including potential cumulative impacts: Renewable energy zones can provide certainty to developers and regulators, and facilitate accelerated permitting (see 4.3.2 and 4.4.1). Such plans should be subject to an SEA (see 5.1.2).
- Improving data quality and accessibility: Sharing data on biodiversity and its interaction with renewable energy across projects and jurisdictions could reduce the burden of data collection in environmental surveys and EIAs, and increase certainty for developers, regulators and stakeholders. Governments can lead on environmental mapping, establish digital data banks and require standardised reporting and disclosure of environmental surveys by developers (Peplinski et al., 2021[36]; EC, 2022[31]; Vasconcelos et al., 2022[29]). See also 5.1.4.
- Developing clear EIA policies and guidelines on EIA, application of the mitigation hierarchy and cumulative impact assessment: Clear policies and guidelines on conducting an EIA and cumulative impact assessment can promote efficient and effective application of EIA, reducing the risk of permitting delays (see 5.1.2).
- Incorporating biodiversity and social criteria into tendering processes for renewable power: Accounting for biodiversity impacts when assessing tenders could help build societal support for projects and reduce the risk of legal action (Energy Transitions Commission, 2023[30]) (see 5.3.1).
- Encouraging developers to adopt and deliver biodiversity-aligned strategies at the company level and in their projects: Demonstrating a commitment to protect and restore nature could increase stakeholder support and reduce conflicts. For example, Iberdrola has set a target of having a net positive impact on biodiversity by 2030. In Spain, Iberdrola (a Spanish power company) set aside EUR 40 million (~ USD 43 million) to protect plant life from 2018-19. In Brazil the company is creating a biodiversity corridor connecting forest and permanent conservation areas (Iberdrola, 2023[37]). Ørsted (a Danish power company) has partnered with the Lincolnshire and Yorkshire Wildlife Trusts to restore biodiversity around the Humber, a large tidal estuary on the east coast of Northern England. The initiative will invest more than DKK 22 million (~ USD 3 million) to restore seagrass and salt marsh and introduce half a million native oysters to improve the health and resilience of the estuary's ecosystem (Orsted, 2022[38]).

5.1.2. Strategic environmental assessment and environmental impact assessment requirements

Strategic environmental assessments (SEAs) and environmental impact assessments (EIAs) are environmental planning and management tools. While EIA is widespread globally, SEA is applied in approximately 40 countries (including all EU Member States) although not per se for renewable power infrastructure (UN Environment, 2018[39]). A review of National Reports to the 13th Conference of the Parties to the Convention on the Conservation of Migratory Species of Wild Animals (CMS) found that 24%

of the 65 countries[2] that reported on renewable energy infrastructure or power lines mentioned conducting SEA for plans or programmes in the energy sector (CMS, 2020[40]).

SEA and EIA are complementary tools to assess the potential environmental impacts resulting from development and incorporate this information into decision-making. SEA is applied at a strategic planning level, for example, to assess the potential impacts of government plans, programmes or policies. EIA is used to assess the potential impacts from projects. Key steps involved in both SEA and EIA include an initial screening, scoping and assessment, consultation and monitoring. Typically, SEA and EIA are triggered when an initial screening indicates potentially significant adverse environmental effects (UN Environment, 2018[39]). The importance of environmental assessment is underscored by the OECD Council Recommendation on the Assessment of Projects, Plans and Programmes with Significant Impact on the Environment (Box 5.5).

Several countries have taken steps to better integrate biodiversity into SEA and EIA (UN Environment, 2018[39]). To promote these efforts, the EU's EIA Directive lists biodiversity as a factor to be considered and has prepared "Guidance on Integrating Climate Change and Biodiversity into Strategic Environmental Assessments" and "Guidance on Integrating Climate Change and Biodiversity into Environmental Impact Assessments". The CBD requests Parties to require EIAs of proposed projects that are likely to have significant adverse effects on biodiversity, with a view to avoiding or minimising such effects (CBD, article 14.1.a). At CBD COP8, Parties endorsed voluntary guidelines on biodiversity-inclusive environmental impact assessment and draft guidance on biodiversity-inclusive strategic environmental assessment (CBD, 2006[41]). At CBD COP11, Parties endorsed voluntary guidelines for the consideration of biodiversity in EIAs and SEAs in marine and coastal areas (CBD, 2012[42]).

Establishing legal requirements and guidance for biodiversity-inclusive SEA and EIA in the power sector – and ensuring their adequate implementation – could be pivotal to ensuring renewable power developments do not compromise efforts to achieve biodiversity goals. This section examines each of these tools in turn, identifying challenges, opportunities, and good practices to scale up their use and effectiveness.

Box 5.5. OECD Council Recommendation of the Council on the Assessment of Projects, Plans and Programmes with Significant Impact on the Environment

The eight recommendations of the OECD Council on the Assessment of Projects, Plans and Programmes with Significant Impact on the Environment are:

1. Use environmental assessment as part of the planning, development and decision-making process for projects, plans and programmes having potentially significant impact on the environment.
2. Establish clear scope and procedures for assessment of the environmental impacts and for determination of relevant mitigation measures as inputs to the planning and decision-making process in order to restore and enhance environmental quality.
3. Incorporate analysis of reasonable alternatives in the assessment of environmental impacts of projects, plans and programmes with a view to arriving at an informed decision that includes best environmental considerations.
4. Include practical and appropriate measures for consulting public authorities having functions and responsibilities relevant to the environmental impacts of projects, plans and programmes.
5. Implement, where appropriate, practical measures for informing the public and for participation by those who may be affected at suitable stages of decision-making on projects, plans and programmes.
6. Ensure that there are means of putting into effect measures derived from the environmental assessment of projects, plans and programmes.
7. Implement appropriate practical measures for monitoring the effects on the environment of projects, plans and programmes that have been subject to environmental assessment.
8. Institute, as appropriate, environmental assessment procedures for projects, plans and programmes that might have significant transboundary impacts.

Source: OECD, Recommendation of the Council on the Assessment of Projects, Plans and Programmes with Significant Impact on the Environment, OECD/LEGAL/0172.

SEA

Considerable scope exists to strengthen the application of SEA. While SEA is applied systematically in some countries, in other countries it is *ad hoc* and sporadic. Of the forty countries with SEA, only a subset has legal requirements to conduct an SEA (UN Environment, 2018[39]). These legal requirements differ considerably in terms of which policies, programmes and plans and which sectors are subject to assessment. The lack of legal force of SEA provisions is cited as a limiting factor in SEA implementation (UN Environment, 2018[39]).

Given the potential for significant, cumulative, impacts on biodiversity from renewable power development, SEA has a clear role in strategic planning in the power sector. SEA can be applied at multiple scales (regional, national, local), depending on the policy, plan or programme. Examples of where SEA could be applied in the low-emissions electricity transition include low-emission long term development plans, climate and energy plans, ten-year grid development plans, broader national and economic social plans, spatial planning and large renewable energy developments. For example, France's multi-annual energy programme was subject to an SEA, which evaluated impacts on biodiversity as well as climate, and the physical and human environment (MTES, n.d.[43]). The SEA listed the positive and negative impacts of the programme's measures on biodiversity and provided mitigation measures for each renewable power technology to respect the mitigation hierarchy.

To optimise their effectiveness, SEAs can be explicitly tiered with EIAs so that they are considered in sequence. The strategic planning level at which SEA is applied is more conducive to assessing cumulative impacts and identifying development scenarios that avoid or minimise these impacts (Josimović, Cvjetić and Furundžić, 2021[44]; Nwanekezie, Noble and Poelzer, 2022[45]). Many of the recurring issues during EIA and project licencing that cause delays and conflicts could be prevented by adequate planning supported by SEA (Dutta et al., 2021[46]). An SEA provides the terms and conditions for project developments and regional monitoring, whereas an EIA assesses in greater detail the potential impacts of different project implementation options.

Given their interdependence, data-sharing across and among SEAs and EIAs is good practice. Data-sharing can help to overcome the data limitations facing environmental assessments and reduce the data-gathering burden on developers (see also 5.1.4). Ideally, the data collated through the spatial planning process and associated SEA would inform project-level assessments by identifying knowledge gaps and highlighting measures that may be required to avoid or minimise significant impacts. Similarly, data collected through project-level EIAs could help to inform spatial planning processes and adaptive management approaches.

SEAs are intended to assess alternative options. However, the adequate assessment of alternative options, though often required by legislation, tends to be lacking (UN Environment, 2018[39]). Developing clear guidelines on how to assess alternatives and establish baselines could facilitate the identification of alternative options (Zhang, Christensen and Kørnøv, 2013[47]). Additionally, initiating an SEA at the earliest stages of planning (i.e. when strategic aims and goals are being set), rather than waiting for a plan, programme or policy to be drafted, could enable a broader range of alternatives to be considered (UN Environment, 2018[39]).

Other good practices include cross-sectoral collaboration to account for cumulative impacts across sectors (see Chapter 4) and engagement of environmental experts, local and indigenous communities and other stakeholders throughout the SEA process. This can help ensure that the most relevant and robust environmental information is considered in decision-making and that the divergent interests and values of stakeholders are accounted for. A critique of SEA in some countries has been the lack of guidance on public participation in SEA and the lack of mechanisms for public participation at the screening or follow-up stage (UN Environment, 2018[39]). Participation is often limited to a mechanism for submitting comments following the publication of documents. Many countries do not ensure public access to SEA monitoring results and evaluations in their national legislation (UN Environment, 2018[39]).

EIA

Environmental impact assessments are widely used to assess potential impacts of development projects, including renewable power infrastructure. Legislative requirements for EIA are in place in most countries, but criteria for when EIAs are required for wind, solar and transmission infrastructure differs (see 5.1.1).

The effectiveness of EIAs varies considerably within and across countries. Shortcomings in implementation can stem from gaps in legislation, a lack of due process or limited capacity to comply with legislation (UN Environment, 2018[39]; Caro-Gonzalez, Toro and Zamorano, 2021[48]). Given targets for rapid renewable power expansion, it would be prudent for countries to review legislation, guidance, governance and institutional arrangements to ensure they facilitate efficient and effective application of EIA for renewable power projects likely to have significant adverse impacts on biodiversity. Countries may find value in developing sector-specific guidance on environmental impact assessments. For example, France developed tailored guidance on environmental assessments for ground-mounted solar energy for public and private developers, EIA practitioners, regulators and other stakeholders (Government of France,

2011[49]). South Africa developed guidance on EIA for renewable energy projects to assist competent authorities and applicants (Deparment of Environmental Affairs, n.d.[50]).

As with SEA, a critical component of EIA is the development of alternatives (i.e., alternative projects or project designs) to identify the approach with the lowest environmental impact. While many countries require the consideration of alternatives, some do not (e.g., Georgia and the Peoples' Republic of China) (UN Environment, 2018[39]). Further, while it is considered good practice to include a "no project" option, this is not widely practiced. Adapting legislation and adopting accompanying guidance on the development of alternatives may be necessary for some countries. The EU, for example, revised its EIA Directive in 2014 making it mandatory to include "a description of the reasonable alternatives studied by the developer" and the reasons for their choice, rather than just "an outline of the main alternatives" as stipulated in the previous version of the Directive.

To maximise their effectiveness, EIAs should consider not only direct impacts but also indirect and cumulative impacts of a project on biodiversity (see Box 5.6 for an example of a cumulative impact assessment). The importance of cumulative impact assessments is increasingly recognised, and it is a mandatory component of EIA in several countries (e.g., Bhutan, Brazil, Canada, EU Members States, Kenya, Panama). However, implementing cumulative assessments in a robust, consistent and cost-effective way remains challenging (Gill and Hein, 2022[51]). For example, offshore wind power projects in the UK during the first two rounds of licensing faced long delays in permitting due to uncertainty about the projects' cumulative impacts (Durning and Broderick, 2018[52]).

To help overcome challenges in assessing cumulative impacts, governments could develop and disseminate guidelines on assessing cumulative impacts. For example, in response to the delays in UK offshore wind power projects, the industry and regulators co-created cumulative impact assessment guidelines, which have been applied by the industry and have helped to improve efficiency and transparency (Durning and Broderick, 2018[52]). Scotland's Nature Agency has published guidance on "assessing the cumulative landscape and visual impact of onshore wind energy developments" (NatureScot, 2021[53]) and on "assessing the cumulative impacts of onshore wind farms on birds" (NatureScot, 2018[54]), which intend to guide cumulative impact assessment during both strategic planning and project development. Several governments have developed general guidance on cumulative impact assessments (e.g., Canada, EU) (UN Environment, 2018[39]), which could be applied to renewable power projects. A technical report prepared by the US National Renewable Energy Laboratory for the IEA's Working Together to Resolve Environmental Effects of Wind Energy (WREN) initiative outlines current practices, challenges and opportunities for assessing cumulative impacts specifically for wind developments (Gill and Hein, 2022[51]).

Assessing cumulative impacts already at the screening stage is good practice (UN Environment, 2018[39]). If cumulative impacts are not considered at the screening stage, it is possible that projects considered to have insignificant direct and indirect impacts are permitted to proceed without an EIA, despite posing considerable risk to biodiversity when accounting for cumulative impacts. The example of the cumulative effects of medium-sized solar energy in Korea and Japan highlighted in Chapter 3, illustrate this concern (Kim et al., 2021[55]).

In addition to stipulating requirements for cumulative impact assessments, EIA legislation could be strengthened through explicit reference to the mitigation hierarchy. Few national EIA laws refer to the mitigation hierarchy, and this is considered a shortcoming that could lead to its *ad hoc* and inconsistent application (UN Environment, 2018[39]). As outlined in Chapter 4, efforts to mitigate biodiversity impacts should place the emphasis first on avoidance measures before considering minimisation, then onsite restoration and offset measures for the residual impacts.

As with SEA, critics point to the lack of adequate mechanisms for stakeholder participation in EIAs (UN Environment, 2018[39]). Ensuring adequate and effective mechanisms for stakeholder engagement throughout the EIA process is essential to inform the EIA and subsequent project decisions. In addition to environmental experts, it is important to engage indigenous and local communities to ensure their rights are respected and their knowledge of local biodiversity is considered. Implementation of the CBD's "Akwé: Kon Voluntary Guidelines on Environmental and Socio-cultural Assessment" could support efforts to ensure the full and effective participation of indigenous and local communities during EIA processes (SCBD, 2004[56]).

Box 5.6. Assessing cumulative impacts of the Tafila Regional Wind Power Project

An example of an *ex-ante* cumulative impact assessment of wind energy is the IFC-commissioned Tafila Region Wind Power Project Cumulative Effects Assessment (CEA) in Jordan. The CEA was conducted at a landscape scale, encompassing the Dana Biosphere Reserve and surrounding Dana Important Bird and Biodiversity Area, and five wind energy sites. The CEA involved three phases:

1. Scoping: This phase provided an initial review of existing data, engaged stakeholders, determined spatial and temporal boundaries of the CEA, and conducted a screening process to select Valued Environmental and Social Components (VECs). Birds, bats and "habitats and other species" were identified as the three VEC categories potentially at risk.
2. Supplementary data and capacity development: Activities included an ornithological training for Jordanian bird surveyors on standardised methods for conducting bird flight activity surveys at wind energy facilities; standardised surveys at wind energy facilities to augment collision risk model (CRM) suitable data, comparable across all the sites; a CRM analysis to obtain species-specific annual fatality rate estimates for at-risk bird populations; compiling of a common database of bird flight activity records from the various sites; a trends analysis using the common database to complement the CRM and allow better understanding of the flight behaviour of Migratory Soaring Birds (MSBs) in the study area; and a reconnaissance site visit and rapid effects assessment for bats and for habitats and other species
3. CEA Framework: A six-step CEA framework using a risk-based approach was developed for birds. The objective was to identify priority VECs at highest risk of cumulative effects so that mitigation, monitoring, and management measures could be put in place to safeguard these populations. The process identified species populations potentially at risk (step 1), evaluated their sensitivity (relative importance and vulnerability) (step 2), and assessed the likelihood of cumulative effect of WPPs on each species population (step 3). Species with the highest risk ratings were determined to be priority VECs. Fatality thresholds were then determined for each priority bird VEC using species-specific demographic information, CRM results, and external stressor fatality estimates from an expert panel (step 4). A site-specific and joint Mitigation and Monitoring Plan (MMP) was then created (step 5), and institutional and information management arrangements proposed (step 6).

Thirteen priority bird species were assessed to be at highest risk through a process that included data analysis, literature review, expert review, and reasoned evaluation of population sensitivity and likelihood of collision risk. Eleven were raptors comprising four migratory soaring bird populations that use the Rift Valley/Red Sea flyway, seven species that are resident or summer breeding raptor populations and two passerine species. Populations of these bird species were assessed to determine an annual threshold of mortalities that each could sustain without affecting their long-term viability. For all 13, the target was determined to be zero mortalities. Mitigation and monitoring measures for birds as well as institutional arrangements were included in the CEA Mitigation and Monitoring Plan (MMP).

Source: (IFC, 2017[57]), Tafila Region Wind Power Projects Cumulative Effects Assessment, www.ifc.org/wps/wcm/connect/62ba7322-8006-4ac4-ab50-f7d0bdd51dcb/CEA+Report+2-16-17+web_w+new+cover.pdf?MOD=AJPERES&CVID=lFczcQe

5.1.3. Prohibiting or mandating specific designs or practices

The risk to biodiversity of renewable power and electricity grid infrastructure depends partly on technology design and the construction and operational practices adopted (see Chapter 3). Governments can steer developers towards technologies and practices that entail relatively lower ecological risk by setting mandatory design or performance standards and prohibiting use of technologies of particular high risk. Examples of biodiversity-related standards for renewable power infrastructure are:

- *Standards and bans to reduce risk of electrocution by power lines:* Some power line designs (referred to as "killer poles") pose significantly higher risk of electrocution than other designs yet are still in operation in many countries (particularly developing countries). Bans and standards could be used to ensure only power lines that pose low risks of electrocution are deployed or in operation (Bern Convention, 2004[58]; Prinsen et al., 2012[59]). Several countries have adopted standards and bans that have driven retrofitting of existing power lines and prohibited new deployment of particularly harmful power line designs (Raptor Protection of Slovakia, 2019[60]). For example, a Spanish Royal Decree in 2008 established measures to protect birds from risk of electrocution and collision on power lines (Spain, 2008[61]). The Decree is to be updated following the adoption of the Spanish Biodiversity Strategic Plan in December 2022. Germany requires newly erected masts and technical components of medium-voltage lines to be constructed in such a way that birds are protected against electric shock, and provided a deadline for retrofitting existing medium-voltage lines that posed a risk (European Commission, 2018[62]).

  Specific design standards could include, for example, requiring a minimum distance between conductors to avoid birds touching two conducting elements at the same time. The EU guidance advises a minimum requirement of 140 cm (European Commission, 2018[62]). In countries with large bird species at risk of electrocution, the distance may need to be greater (Sielicki, 2020[63]).
- *Wind turbine operational curtailment:* Operational curtailment such as shutting down turbines during bird and bat migrations, increasing the cut-in speed of wind turbines and feathering (pitching blades parallel to the wind to minimise rotation) have been shown to effectively reduce mortality from collision (see 3.1.2). For example, Whitby, Shirmacher and Frick (2021[64]) estimate that setting the cut-in speed to 5 metres/second (m/s) could reduce the total bat mortality at individual facilities in any given year by 33%–79%. Governments can set requirements to adopt operational curtailment under certain conditions. The US states of Maine and Vermont, for example, legally require all wind farms to increase their cut-in speed. In Canada, Ontario and Alberta require shutdown of certain wind turbines when mortality thresholds are reached (Le Maître, J et al., 2017[65]).

  Governments should seek to adopt curtailment requirements that are both environmentally-effective and cost-effective. Operational curtailment may result in losses in annual energy production. Simulations in the US, for example, estimate annual energy production loss of <1% to >10% depending on the curtailment scenario. Curtailment strategies based on weather, season and time of day would minimise risks at lower cost than blanket curtailment (Squires et al., 2021[66]). Smart curtailment approaches that leverage technology to provide and respond to real-time data may facilitate more cost-effective mitigation (Hayes et al., 2019[67]; McClure et al., 2022[68]; Sheppard et al., 2015[69]).
- *Standards for regulating piling noise from offshore wind turbine construction:* Pile-driving is a source of disturbance for marine wildlife such as harbour porpoises (see Chapter 3), and is regulated by standards in some countries. In Denmark, for example, the accumulated Sound Exposure Level ($SELC24h$) from each piling sequence must not exceed a threshold value of 190 $dB\ re$ 1 $\mu Pa2\ s$ (porpoises permanent threshold shift). If the estimated $SELC$ exceeds the threshold the source level must be mitigated accordingly. If the actual accumulated SEL exceeds the threshold value, then the concessionaire must take measures to identify the causes of this

deviation and perform corrective measures, including adjusting the installation method. The Danish guidelines are expected to evolve as marine bioacoustics develop so that more species-specific weightings can be used, accounting for the specific hearing sensitivities of each species when estimating the impact of a given noise source. Species-specific weightings could both reduce over-regulation and ensure that harm to all exposed species is adequately mitigated (Tougaard, Beedholm and Madsen, 2022[70]).

- *Standards for rooftop solar on new buildings:* Capitalising on the potential of rooftop solar could help accelerate the transition to low-emissions electricity with minimum risk to biodiversity, as it does not entail land or sea-use change (except from mining for mineral components). As part of building standards, governments can mandate new buildings and existing public buildings to have solar arrays. The 2022 Energy Code in California, for example, requires all newly constructed commercial buildings to have a solar array and an energy storage system installed (California Energy Commission, 2022[71]).

To ensure environmental and cost-effectiveness, standards need to be sufficiently stringent and enforced, while also providing the necessary flexibility to foster innovation. For example, various solutions exist for reducing piling noise, which may vary in cost-effectiveness and feasibility in different contexts. By setting scientifically based noise restrictions on piling, rather than prescribing a technology, governments can foster innovation, encouraging industry to develop cost-effective noise mitigation solutions. Standards may also need to be regularly updated to reflect latest knowledge of renewable energy infrastructure impacts and mitigation effectiveness, and to provide incentives for continuous innovation.

5.1.4. Monitoring, data sharing and disclosure requirements

Post-construction monitoring of renewable energy projects is fundamental for ensuring renewable power expansion is biodiversity-aligned. Post-construction monitoring serves several purposes (European Commission, 2020[19]). First, it tests the validity of the conclusions, both in the short and long-term, arising from environmental impact assessments (European Commission, 2020[19]). Such validation is particularly important considering uncertainty around tipping points, cumulative impacts, the effects climate change will have on biodiversity and ecosystem services and the accuracy of models. For example, post-construction monitoring of bird or bat mortality at wind farms can assess whether predictions from collision risk models were accurate. Second, monitoring can ensure mitigation measures adopted are effective throughout the project's lifetime. Information gleaned from post-construction monitoring allows developers to adjust their projects to ensure biodiversity is safeguarded.

Project-level monitoring can also provide broader benefits, by informing other projects and strategic landscape or seascape level planning. Monitoring data can contribute to the knowledge base on context-specific biodiversity impacts from renewable energy and provide insights on the effectiveness of mitigation measures. The data can inform environmental impact assessments – including cumulative impact assessments – permitting requirements and decisions, and ultimately the location and design of renewable energy projects.

Governments can mandate post-construction monitoring of renewable energy impacts (e.g., as a condition for obtaining an environmental permit). Additionally, governments could develop or promote protocols or guidelines for monitoring the biodiversity impacts of renewable energy projects. By stipulating specific monitoring approaches or requirements, protocols can ensure monitoring is adequate and standardised (while allowing flexibility to ensure the most appropriate monitoring approaches are adopted for the habitats or species of concern), thereby allowing comparison of impacts and analysis of combined impacts from multiple projects. Monitoring guidelines could outline principles and good practices for developing indicators, selecting methodologies, defining the appropriate spatial scale, monitoring effort, timing and frequency, and standardised data collection to facilitate data sharing.

Several governments require monitoring of renewable power projects (e.g. Germany), have developed protocols for monitoring renewable energy projects (e.g. France, Box 5.7), or have produced guidelines (e.g. New York State, US; US east coast; Saskatchewan, Canada; South Africa) (Aronson et al., 2020[72]; Jenkins et al., 2015[73]) (New York State, 2016[74]; MoE Saskatchewan, 2018[75]) (ROSA, 2021[76]). However, many countries developing renewable power do not have such protocols or guidelines. Furthermore, a review of monitoring guidance and protocols for offshore wind in the Baltic Sea and North Sea found inconsistencies in how methodologies are applied and recommended greater consultation and harmonisation of methods at an international and regional level (Stephenson, 2021[77]).

Box 5.7. France's monitoring protocol for wind energy developments

The Ministry for Ecological Transition in France has developed national guidance for implementing monitoring of wind energy development projects in relation to birds and bats. The main objectives are to:

- Assess the real effects (bird and bat collisions) and the effectiveness of mitigation measures;
- Obtain sufficient data from several wind farms to calculate average mortality rates for birds and bats;
- Collect data at national level to underpin future policy and actions.

This protocol requires at least one post-construction monitoring measurement during the first 3 years of operation. If no significant effects are identified, at least one follow-up measurement should take place in the next 10 years. If significant effects are observed, corrective measures must be implemented and a new post-construction monitoring measurement must be carried out within the next year. The protocol gives precise instructions on the periods of the year when monitoring must be conducted. These periods should always be relevant for the specific case. For example, some wind farms might have more effects on wintering waterfowl, while other wind farms might have more effects on breeding raptors. The protocol also gives precise instructions on: (i) the number of countings (at least 20); (ii) the number of turbines that must be monitored; (iii) the method for searching for carcasses, etc. For bats, the monitoring campaign must measure in pre-defined periods (specified in the protocol) both bat activity at the level of the turbine and carcasses on the ground.

Source: (MTES, 2018[78]), Environmental monitoring protocol for terrestrial wind farms.

While the specific approach and focus of monitoring may differ across wind, solar, power lines and other infrastructure, good practices generally include:

- Defining clear objectives and scope for monitoring. Monitoring should be designed to respond to clearly defined questions, thereby providing the necessary information to understand and mitigate adverse impacts from renewable energy projects. Extensive but poorly defined monitoring programmes risk being "data-rich, information-poor" and unnecessarily costly (Wilding et al., 2017[79]). It is also important that monitoring effort is commensurate with the risk posed by the project (Bennun et al., 2021[80]).
- Adopting indicators based on the Pressure-State-Response model, and ensuring they are relevant (see point above), analytically sound and measurable (OECD, 1993[81]). For a wind farm, for example, a pressure indicator could be "fatal avian collisions with wind turbines", a state indicator could be "population size of affected avian species" and a response indicator could be "number of wind turbine shutdowns-on-demand" (Bennun et al., 2021[80]).
- Establishing a pre-project baseline against which to assess any changes. A common approach for pre and post-construction monitoring is the before-after-control-impact (BACI) model. In addition

to monitoring the baseline and changes at the project site it may be necessary to monitor a control site to account for any background environmental variability (European Commission, 2020[19]).

- Standardising the indicators and methodology for biodiversity surveys to ensure comparability of results pre- and post-construction and throughout the lifetime of the project (European Commission, 2020[19]).
- Matching timing and frequency of monitoring efforts to temporal movements of species and phenology (Amerson et al., 2022[82]). For example, monitoring impacts of a renewable power facility during breeding seasons and migration will be important for some projects.
- Coordinating monitoring schemes across projects and landscapes to track and mitigate cumulative impacts more effectively. Belgium, for example, has established a monitoring programme for offshore wind projects. Monitoring is required in the environmental permit and coordinated by the Operational Directorate Natural Environment of the Royal Belgian Institute of Natural Sciences (RBINS, 2023[83]).

To ensure project monitoring delivers wider benefits for renewable energy development, monitoring data must be robust and easily accessible for other developers and energy planners. While some developers already share their data publicly, such as the Wolfe Island Wind Farm in Ontario, Canada (Transalta, 2022[84]) and the Gullen Range Wind Farm in New South Wales, Australia (Gullen Range Wind Farm, 2016[85]; Bennun et al., 2021[80]), many do not (Stephenson, 2021[77]). Significant scope exists to improve the quality and accessibility of monitoring data and enhance transparency. Governments have a role in driving improvements in data quality and transparency, for example, by encouraging or requiring developers to report data and insights from their pre- and post-construction monitoring and providing platforms to facilitate data sharing (Box 5.8).

The momentum building around nature-based disclosure, through initiatives such as the Taskforce Nature-Related Financial Disclosure, could also play a role in addressing data and transparency challenges. Specifically, nature-based disclosure could help shift investments towards renewable power projects that are environmentally-sound and away from more harmful developments. Several jurisdictions have already introduced regulations that require nature-related disclosure in some contexts (e.g. Article 29 of the French law on Energy and Climate (France Treasury, 2021[86]) and the EU Sustainable Finance Disclosure Regulation (EU, 2022[87])).

Box 5.8. Improving data quality, access and sharing

Limitations in the availability, quality and application of environmental data are widely reported challenges that undermine the effectiveness and efficiency of SEA, EIA and permitting processes for renewable power projects. Data challenges can result in negative biodiversity impacts and economic costs. For example, the environmental assessment of the London Array wind farm in the UK underestimated potential impacts on Red-throated Divers (*Gavia stellata*) and Little Auks (*Alle alle*) due to data deficiencies. Significant population density declines were observed in proximity to the wind turbines resulting in the second phase expansion, which spanned 39km^2 and comprised 56 new turbines, being cancelled late in the planning phase.

Governments can support improvements in data availability and quality by developing or funding central data repositories, biodiversity mapping tools and guidelines on data access and use. In the EU, the INSPIRE Directive has promoted efforts to increase public access to spatial data, including on species, habitats and energy resources, and to coordinate data use across governments. Several governments have developed centralised portals that collate environmental assessments or integrate biodiversity data that can be used to inform SEA/EIA and subsequent mitigation measures. For example:

- Brazil has established an online repository of environmental impact studies to improve knowledge and data, including on renewable energy impacts.
- Denmark's Natural Environment Portal provides access to a variety of environmental data that can inform SEA and EIA.
- Ireland's Environmental Protection Agency maintains an up-to-date SEA Spatial Information Sources Inventory, which covers biodiversity and other environmental assets.
- European Union's European Marine Observation and Data Network (EDMODnet) processes marine data according to international standards and make that information freely available as interoperable data layers and data products.
- The Crown Estate's Marine Data Exchange in the UK collates marine industry survey data, research and evidence. It provides access to information collected during pre- and post-construction phases of offshore wind energy projects with the aim of de-risking investments and reducing survey costs.

Legal requirements could help ensure data collected through SEA/EIA processes and post-construction monitoring are collected, reported and managed consistently (i.e., standardised), and made publicly accessible. For example, New York State law requires offshore wind energy developers selling power to New York State to make non-proprietary environmental data publicly available "as soon after collection [as] is practicable for use by third parties in decision-making around adaptive management". In addition to supporting adaptive management, this requirement is intended to help fill knowledge gaps concerning marine wildlife populations and ecosystem dynamics given the projected growth in offshore wind.

Source: (Brazil, 2022[88]) (COWI, 2009[89]) Study concerning the report on the application and effectiveness of the EIA Directive Final report; (Howard, 2018[90]) Industry evidence programme: offshore windfarms - pilot industry evidence base. Report to The Crown Estate and Royal HaskoningDHV; (Ireland EPA, 2022[91]), SEA Spatial Information Sources Inventory, Kettel el al., (2022[92]), Better utilisation and transparency of bird data collected by powerline companies; (The Crown Estate, 2023[93]), Marine Data Exchange; Underwood et al. (2018[94]), The use of biodiversity data in spatial planning and impact assessment in Europe;. (NYSERD, 2021[95])Wildlife Data Standardization and Sharing: Environmental Data Transparency for New York State Offshore Wind Energy.

5.1.5. Due diligence and responsible business conduct requirements

Responsible Business Conduct (RBC) due diligence is a process for financial and non-financial companies to identify, prevent, mitigate and account for their actual and potential adverse impacts on the environment (including biodiversity) and other RBC issues (e.g. human rights, labour rights, bribery and corruption) (OECD, 2018[96]). These impacts may arise in a company's own operations, supply chain and other business relationships. Owing to the potential adverse impacts of renewable power on nature and people, adopting a robust risk-based due diligence approach is good practice for renewable energy companies. The above-mentioned instruments, EIAs and SEAs, can be fully integrated as part of an RBC due diligence process to identify potential risks and impacts on biodiversity of renewable power infrastructure projects (step 2 of the due diligence process). More specifically, integration of RBC-related due diligence in the context of project and asset finance can help investors, including development finance institutions, in identifying, preventing and mitigating biodiversity-related risks in the project and asset they finance, including in renewable energy infrastructure projects. To that end, the OECD published recommendations on how to conduct RBC due diligence in the context of project and asset finance transaction, creating baseline expectations amongst other international standards such as the IFC Performance Standards and Equator Principles (Box 5.9) (OECD, 2022[97]).

While due diligence and responsible business conduct is voluntary in most jurisdictions, an increasing number of governments are adopting laws requiring companies to undertake human rights and environmental due diligence. For example, France's 2017 Duty of Vigilance Law requires large French companies to develop a due diligence plan to identify and prevent adverse impacts on the environment among other things, arising directly or indirectly from the operations of the company and companies it controls. Due diligence laws have since been adopted in Germany and Norway (2021), while the European Commission has adopted a proposal for a Directive on corporate sustainability due diligence (2022). Large renewable power companies fall under this legislation, whereas small and medium enterprises are excluded.

OECD's Due Diligence Guidance for Responsible Business Conduct (OECD, 2018[96]) helps companies to understand and implement due diligence for RBC as foreseen in the OECD Guidelines for Multinational Enterprises (OECD, 2011[98]). The Guidance also seeks to promote a common understanding amongst governments and stakeholders on due diligence for RBC. Typically, a company considers risk to themselves (e.g., financial risk, market risk, operational risk, reputational risk), however the guidance takes an outward-facing approach to risk, referring to the likelihood of adverse impacts on people, the environment and society that enterprises cause.

Box 5.9. The International Finance Corporation's Environmental and Social Performance Standards and the Equator Principles

IFC Performance Standards

The IFC Performance Standards provide guidance to IFC clients on how to identify risks and impacts. They are designed to help avoid, mitigate, and manage risks and impacts as a way of doing business in a sustainable way. In the case of its direct investments (including project and corporate finance provided through financial intermediaries), IFC requires its clients to apply the Performance Standards to manage environmental and social risks and impacts so that development opportunities are enhanced. Together, the eight Performance Standards (PS) establish standards that the client is to meet throughout the life of an investment by IFC.

PS 6: Biodiversity Conservation and Sustainable Management of Living Natural Resources has the following objectives: i) To protect and conserve biodiversity; ii) To maintain the benefits from ecosystem services; iii) To promote the sustainable management of living natural resources through the adoption of practices that integrate conservation needs and development priorities. The applicability of this PS is established during the environmental and social risks and impacts identification process. Based on the risks and impacts identification process, the requirements of PS6 are applied to projects i) located in modified, natural, and critical habitats; ii) that potentially impact on or are dependent on ecosystem services over which the client has direct management control or significant influence; or iii) that include the production of living natural resources. Notably, in areas of critical habitat, the client will not implement any project activities unless all of the following are demonstrated:

- No other viable alternatives within the region exist for development of the project on modified or natural habitats that are not critical;
- The project does not lead to measurable adverse impacts on those biodiversity values for which the critical habitat was designated, and on the ecological processes supporting those biodiversity values;
- The project does not lead to a net reduction in the global and/or national/regional population of any Critically Endangered or Endangered species over a reasonable period of time;
- A robust, appropriately designed, and long-term biodiversity monitoring and evaluation programme is integrated into the client's management programme.

Equator Principles

The Equator Principles provide a common baseline and framework for financial institutions to identify, assess and manage environmental and social risks when financing projects. They apply globally to all industry sectors (including solar, wind and electricity network projects) and five finance instruments: 1) project finance advisory services; 2) project finance; 3) project-related corporate loans; 4) bridge loans; and 5) project-related refinance and acquisition finance. Currently, 137 financial institutions in 38 countries have voluntarily adopted the ten Equator Principles. By implementing the Equator Principles, negative impacts of projects on biodiversity and communities should be avoided where possible. Unavoidable negative impacts should be reduced, mitigated and/or compensated for. The Equator Principles draw on the International Finance Corporation's (IFC) Environmental and Social Performance Standards.

Source: (IFC, 2012[99]), International Finance Corporation's Environmental and Social Performance Standards (Equator Principles Limited, 2022[100]), The Equator Principles, https://equator-principles.com/about-the-equator-principles/.

The OECD Guidelines for MNEs (hereafter "Guidelines") have a unique promotion and grievance mechanism – the National Contact Points (NCPs). Examples of the OECD's Due Diligence Guidance grievance mechanism being employed for renewable energy and transmission infrastructure are outlined below:

- In October 2012, the Swedish and Norwegian NCPs received a submission from the Sami reindeer herding collective in Jijnjevaerie Sami Village alleging that Statkraft AS, a Norwegian multinational business, had not observed the general policies, human rights, and environment provisions of the Guidelines by planning to build a wind power plant on reindeer herding ground in Sweden. The Sami reindeer herding contended that "meaningful engagement" had not taken place and requested that the NCPs mediate between the parties to reach a positive solution. Mediation took place but no agreement could be reached. The NCPs did not find any grounds for concluding that Statkraft had failed to comply with the OECD Guidelines. However, they identified areas for improvement. The NCPs recommended that the parties showed renewed will to negotiate an agreement on the further development of the wind power projects, their scope and extent and compensation schemes. The affected parties reached an agreement on their own following the conclusion of the NCP process (OECD, 2012[101]).
- In December 2017, Stichting Hou Friesland Mooi (HFM) submitted a specific instance to the Dutch National Contact Point (NCP) concerning an alleged violation of the OECD Guidelines for Multinational Enterprises by Nuon Energy N.V. and/or Nuon Wind Development B.V. (Nuon). At the advice of the NCP a dialogue took place between the notifying party and company. During the second meeting on 31 August 2018, the parties reached agreement on several points, and indicated that the NCP dialogue had been useful in clarifying the points presented in the notification. The parties agreed that in the new phase of the project, consultation with the local community would begin anew and the parties would aim to restore confidence. On December 18, 2018, a final statement was published. The NCP concluded that there was a lack of clarity regarding Nuon's role in relation to the provincial authority when concerned with the engagement of the project's stakeholders, and that Nuon is obliged to comply with the provisions of the OECD Guidelines on its own initiative. Additionally, the NCP recommended that Nuon should communicate more clearly and publicly what its role is in relation to the provincial authority and local community. Finally, the NCP recommended that the parties continue their dialogue (OECD, 2017[102]).

5.2. Economic instruments

Biodiversity-related economic instruments can provide continuous incentives to both producers and consumers to behave in more environmentally sustainable ways. By raising the cost of activities that harm or degrade biodiversity (e.g., taxes; biodiversity offsets) and rewarding activities that benefit biodiversity (e.g., subsidies; payments for ecosystem services), economic incentives encourage producers and consumers to behave more sustainably. Economic instruments used to mainstream biodiversity into renewable energy infrastructure development include biodiversity offsets and environmentally motivated subsidies.

5.2.1. Biodiversity offsets

Biodiversity offsets are measurable conservation outcomes that result from actions designed to compensate for significant, residual biodiversity loss resulting from development projects (OECD, 2016[103]). They aim to internalise the external costs of development by imposing a cost on the activities that cause biodiversity loss and are therefore based on the polluter pays approach. Offsetting is the last step in the mitigation hierarchy (see Chapter 3); offsets are intended to be implemented only after all reasonable steps have been taken to avoid and minimise biodiversity loss at the development site and restore on-site impacts. Effectively designed offsets within a mitigation hierarchy framework can be in instrumental in achieving objectives of "no net loss" or "net gain".

Three key approaches exist for implementing a biodiversity offset (OECD, 2016[103]): one-off offsets, biobanking and in-lieu payments. One-off offsets are undertaken by the developer or by a third-party provider on their behalf. Biobanking involves a repository of offset credits, each which represent a quantified gain in biodiversity resulting from actions to restore, enhance, and preserve biodiversity. A biobank is established in anticipation of future development impacts and tends to provide offsets (credits) for multiple projects. Payments-in-lieu are a mechanism by which regulatory agencies levy fees on developers for their adverse impacts on biodiversity. The collected fees are then spent by government agencies or a third-party on compensatory biodiversity measures. The payment is typically based upon a reasonable cost estimate of the financial resources necessary to compensate for the biodiversity loss. Each approach has its advantages and disadvantages (Table 5.2).

Table 5.2. Advantages and disadvantages of the three biodiversity offset approaches

	Advantage	Disadvantage
One-off	Flexibility to deal in nuance way with project-specific impacts	Flexibility may come at expense of consistency and transparency Temporal loss of biodiversity possible as offset typically comes at or around same time as biodiversity loss at the development site, yet offset can take time to mature
Biobanking	Offsets established before project impacts, reducing risk of temporal loss or non-attainment of objectives Tend to be larger in size because serve multiple projects	May not be appropriate where offset demand is low Less flexibility to address nuances of project-specific impacts
In-lieu	Flexibility and simplicity	Disconnect between biodiversity loss and compensation Expenditure of levied fees often not made in timely manner leading to temporal loss of biodiversity

Source: Based on (OECD, 2016[103]), Biodiversity Offsets: Effective Design and Implementation, OECD Publishing, Paris, https://doi.org/10.1787/9789264222519-en and (Bennett, Gallant and Ten Kate, 2017[104]), State of Biodiversity Mitigation 2017.

Two main types of biodiversity offset measures exist: restorative measures and averted loss measures (IFC, 2012[105]). Restorative measures aim to benefit biodiversity by improving the state of habitats or ecosystems outside the project area that have been previously damaged. Averted loss measures aim to reduce pressures on existing biodiversity at an area demonstrated to be under threat of imminent or projected loss. A combination of the two may be required to effectively offset the impacts of some renewable energy projects (see Table 5.3 for renewable energy related examples).

Table 5.3. Examples of types of offset approaches for solar and wind projects

Offset Type	Solar	Onshore wind	Offshore wind
Restoration	Restore degraded areas of similar habitat	Improve condition of preferred raptor habitat Captive breeding and successful reintroduction of raptor species where populations are depleted	Protect and restore prey species stocks Eradicate invasive species from nesting grounds of seabird species Improve condition of foraging or breeding grounds for marine mammals
Avoided loss	Protect a threatened area of similar habitat off-site	Retrofit non-project power lines to prevent bird electrocution or collisions Protect roosts at risk elsewhere for priority bat species Reduce predator-livestock conflict to prevent incidental poisoning of scavenging bird species Protect key stopover, passage, nesting or wintering sites for migratory birds Support awareness, enforcement and alternative livelihoods programmes to reduce illegal capture/hunting of migratory bird species	Protect nesting grounds for migratory birds in their breeding areas (off-site) Support implementation of locally managed marine areas to protect priority species or habitat Support prevention of fisheries bycatch for priority species

Source: (Bennun et al., 2021[80]), Mitigating biodiversity impacts associated with solar and wind energy development: guidelines for project developers, 10.2305/iucn.ch.2021.04.en.

Governments can mandate the use biodiversity offsets as part of project permitting or provide legislation that facilitates the use of voluntary biodiversity offsets. At least thirty-seven countries require biodiversity offsets for infrastructure projects in some contexts[3] (IUCN, TBC and DICE, 2021[106]). Subnational governments have also developed legislation or programmes for biodiversity offsetting specific for infrastructure. For example, New South Wales, Australia, has developed an offset policy for state significant development and infrastructure, which has triggered offsets for solar (e.g., AG Nyngan Solar Park) and wind (e.g., Rye Park Wind Farm) projects. The Scottish Borders Council, Scotland, has developed a biodiversity offsetting programme specifically to offset the impacts of wind energy (Butterworth et al., 2019[107]).

In addition to government policy, biodiversity offsets may be triggered through lending requirements of international financial institutions, such as the International Finance Corporation or implemented voluntarily by developers for example to achieve internal corporate commitments of no net loss or net positive gain. Documented examples of offsets applied to solar and wind exist in several geographic, political and environmental contexts:

- Apennine Wind Farms, Italy: Offset measures were adopted to compensate for the impacts of two wind farms in the province of Macerata, Apennine Mountains. The Biological Territorial Capacity (BTC) indices and ecological energy balance considerations were used to determine residual negative impacts on priority grassland habitat, bats, raptors, and other important bird species. The energy requirements of raptors and the effect of habitat loss due to the wind farms on the raptors' prey sources were used to calculate the area of habitat and associated prey needed to compensate for that loss. Unused and degraded agriculture areas within the Natura 2000 site were then restored to compensate for loss of grassland habitat, human hunting was excluded from an area commensurate with that lost to raptors through the development, and an existing electricity transmission line was buried to reduce risk of bird collisions. Where predictions of impacts were uncertain, relatively larger compensation measures were taken (BBOP, 2009[108]).

- Broken Hill Solar Plant, New South Wales, Australia: A permitting condition for the Broken Hill Solar Plant in New South Wales, Australia, was to develop an Offset Management Package (ngh environmental, 2013[109]). A biodiversity offset management plan was developed for a site located 1.5 km from the development site. The objectives of the plan were to provide a "like for like" offset regarding vegetation types and threatened species habitats, ensure consistency with the Principles for the Use of Biodiversity Offsets in NSW[4] and achieve a net improvement in biodiversity values within the offset site sustained in the long-term. Approximately 150 hectares of vegetation was required to be cleared for the development, covering four vegetation types, including near threatened Black Bluebush and Mulga Woodland. Following desktop research and field surveys, an offset site of 159 hectares was identified at 1.5 km from the development site. Overall, the proposed offset presented a 1:1.1 area impacted to area offset ratio, with a 1:1.3 ratio for Black Bluebush low open shrubland, which was the main vegetation type to be impacted and considered 'near threatened'. Measures included weed control, feral cat and rabbit control, exclusion of feral goats, implementation of controlled burns to reintroduce a more natural fire regime and assisted recovery of degraded areas. Consistent with the Conditions of Approval, a biodiversity offset monitoring plan was developed and results reported annually to the NSW Office of Environment and Heritage (Tebb, 2018[110]).
- The Kipeto Wind Power Project, Kenya: The proposed wind farm of 100 MW comprises 60 wind farms. It is near nesting colonies of two Critically Endangered vulture species: Rüppell's vulture *(G. rueppelli)* and white-backed vulture *(G. africanus)*. Both species regularly fly over the wind farm, but this issue was identified too late in the planning process to avoid impacts through re-siting (Bennun et al., 2021[80]). Instead, on-site monitoring was conducted to help quantify the risks to vultures. Minimisation and offset measures were then developed with an aim of achieving net gain for both species in line with IFC Performance Standard 6. Minimisation measures included rapid detection and removal of carcasses from the site to avoid attracting vultures, and observer-led shut-down-on-demand when birds at risk are spotted. The biodiversity offset for the project includes several measures to address human-wildlife conflict with an objective of reducing retaliatory poisoning of wildlife, which can be lethal to the vultures that feed on the carcasses of poisoned animals. Offset activities are implemented by a partnership of four conservation NGOs and the Kenya Wildlife Service and overseen by a multi-stakeholder Biodiversity Committee (Bennun et al., 2021[80]).
- The Falkenhöhe wind farm project in the Black Forest: The project conducted an impact study (*UVPBericht*), a specific study on the mitigation of impacts (*Eingriffsregelung*) for biodiversity in general (*Landschaftspflegerischer Begleitplan*) and a study specific to protected species (*spezielle artenschutzrechtliche Prüfung*) (Bas and Dieckhoff, 2021[111]). The studies, carried out prior to the authorisation of the project, mention compensatory measures that need to be taken on and off-site. Compensatory measures carried out on the project site included planting trees of local species and restoration of soil functions. Offset measures included improving the habitat for different species (e.g., capercaillie, greater murin) over 24 ha of forest located 8 km from the wind farm, installing 40 bat nest boxes and improving the honey buzzard feeding habitat. According to the regulation of protected species, the offset measures were to be conducted prior to project impacts occurring, be ecologically equivalent and functionally close to the impacted site. The biodiversity loss and gains were quantified using the "eco-points" system, which is a grading system that includes functionally descriptive indicators, i.e. succession stage, degree of nativeness, structural richness, diversity of

species and normative indicators (including rarity of habitat, rarity of species, sensitivity and unfavourable tendency of endangerment) (OECD, 2016[103]). These indicators are used to classify land use and habitat types into 11 categories. Each category is further divided into sub-classifications, to which a certain number of eco-points are assigned per square metre. The benefits of the planned offset measures (i.e. eco-point gains) are expected to be greater than the negative impacts of the project (i.e. eco-point losses) (Bas and Dieckhoff, 2021[111]).

Marine offsets appear to be largely absent for renewable power developments (Vaissière et al., 2014[112]; 2021[113]), but could become increasingly relevant as fixed and floating offshore wind expands. However, policy and practice for marine biodiversity offsets in general is significantly less developed than for terrestrial biodiversity offsets. An analysis of marine biodiversity offset policy identified six countries with policy frameworks for marine biodiversity offsets (Australia, Canada, Colombia, France, Germany and US) and identified seven countries where marine biodiversity principles were being applied outside public policy frameworks (Niner et al., 2017[114]).[5] Another analysis indicated that almost 80 countries have policies that would allow offsetting of marine impacts (Shumway et al., 2018[115]). In practice, marine offsets are seldom conducted (Bull and Strange, 2018[116]) and their efficacy is scarce and patchy (Jacob et al., 2020[117]).

Several challenges for biodiversity offsetting are specific to – or accentuated – in the marine context. These include biophysical challenges (e.g. greater connectivity of marine areas, lower likelihood of restoration success, data paucity) and social or governance issues (e.g. lack of private ownership and greater probability of leakage) (Shumway et al., 2018[115]; Niner et al., 2021[113]). Priorities for further exploring and developing offset approaches in the marine environment include (Jacob et al., 2020[117]; Vaissière et al., 2014[112]; Hooper, Austen and Lannin, 2021[118]): improving marine data; applying EIA thoroughly, including assessment of cumulative impacts of energy development; developing appropriate metrics for offsetting; strengthening understanding of ecological restoration techniques for marine ecosystems; and adjusting policy and planning frameworks. Spatial conservation planning could allow offsets to be pooled at appropriate scales (e.g. regional scale), rather than siloed, project-by-project approaches, which could deliver greater biodiversity benefits (Jacob et al., 2020[117]; Hooper, Austen and Lannin, 2021[118]; Croll et al., 2022[119]).

Biodiversity offsets remain contentious (May, Hobbs and Valentine, 2017[120]), but their effectiveness heavily depends on how they are designed and applied (see Box 5.10). For example, offsets can be counterproductive if the mitigation hierarchy is not respected (Maron et al., 2015[121]; Primmer et al., 2019[122]). However, only 10 of 37 countries where offsetting is mandatory require robust application of the mitigation hierarchy (IUCN, TBC and DICE, 2021[106]).[6] Strengthening regulations on mitigation hierarchy and ensuring compliance with regulations is an important step, which could be usefully accompanied by the development or dissemination of guidelines on implementing the mitigation hierarchy (see e.g. (CSBI, 2015[123])).

Box 5.10. Biodiversity offset design and implementation considerations

Thresholds and coverage

Biodiversity offsets will not always be able to deliver equivalent outcomes because biodiversity may be of exceptional high value, irreplaceable or vulnerable. Establishing thresholds for what can and cannot be offset is therefore key. Coverage refers to the type of biodiversity intended to be addressed (e.g., habitats, species, ecosystem services) and the sectors that are included in the programme (e.g., mining, wind power, property development, agriculture).

Equivalence

As no two sites are ecologically identical, designing offsets requires assessment of how to achieve biodiversity benefits at the offset site that are ecologically equivalent to losses at the impact site. This is known as "like-for-like". Determining ecological equivalence necessitates a comparison of the biodiversity loss and offset sites in three dimensions: biodiversity type, location and time. Some stakeholders may opt for trading up with offsets, known as "like-for-better", by targeting different biodiversity features of higher priority than those impacted.

Additionality

The biodiversity improvements at offset sites should provide new contributions to biodiversity conservation over and above the existing levels. A reference scenario is therefore needed. Biodiversity offsets variously consider protection, restoration, recreation and enhancement measures as additional.

Permanence

Biodiversity offsets should deliver conservation outcomes for at least as long as the biodiversity loss persists at the development site. Land tenure, financial sustainability and appropriate incentives for land management are important components of delivering permanent outcomes.

Monitoring, reporting and verification

Robust MRV methodologies that can assess progress toward an offset's objectives are critical. This includes adequate documentation of management plans, regular monitoring including on-site checks, clear and transparent reporting, and verification by a third party.

Compliance and enforcement

MRV frameworks must be supported by appropriate compliance and enforcement measures to create the incentives necessary for offset suppliers to deliver conservation outcomes over time.

Transaction costs

Transaction costs in offset programmes include costs associated with identifying, creating and securing an offset, applying for development permission, and undertaking MRV and enforcement. Reducing these administrative and time costs will increase the efficiency of an offset programme. Biobanks, for example, reduce the search costs of finding appropriate offset sites for developers.

Source: (OECD, 2016[103]), Biodiversity Offsets: Effective Design and Implementation, OECD Publishing, Paris https://dx.doi.org/10.1787/9789264222519-en.

Onshore and offshore renewable power developments may lend themselves to specific offsetting approaches. For example, as developments are often concentrated in areas of high resource potential (e.g., wind in Gulf of Suez; solar in Mojave Desert, California) potentially affecting similar habitats, aggregated offsets whereby developers pool resources into a joint intervention could reduce transaction costs and improve the effectiveness of offsets (provided there is adequate coordination and governance). While there is limited experience in aggregated offsets, they may become increasingly attractive as countries establish regulatory schemes requiring developers to contribute to specific quantitative conservation targets (Bennun et al., 2021[80]).

Another issue is how to offset the impact on migratory birds, bats and marine life. Impacts on these taxa are common and often the most significant impact from wind energy facilities. Generally, offsets take place in the same jurisdiction as the impacts and are expected to be geographically close. However, for migratory birds, bats and marine life, little benefit may come from trying to offset impacts near the development. The most ecologically appropriate site for offsetting the impacts may be in other countries, for example, at breeding sites or over-wintering grounds. (Bennun et al., 2021[80]) suggest that an international offsetting mechanism could play a role in addressing this issue (e.g., under the CMS).

A variety of biodiversity offsetting tools and metrics have been developed over the past two decades (see (OECD, 2016[103]) for examples). These continue to evolve and are being supplemented by new approaches, some of which have been developed specifically with renewable power development in mind e.g.

- *Avian-Impact Offset Method (AIOM)* (Shaffer, Loesch and Buhl, 2019[124]): the AIOM was developed by scientists from Northern Prairie Wildlife Research Center and the US Fish and Wildlife Service to quantify the amount of habitat needed to provide equivalent biological value for birds displaced by energy infrastructure in the US. The method is based on five metrics: impact distance, impact area, pre-impact density, percent displacement, and offset density. The authors calculated percent displacement values for breeding waterfowl and grassland birds. This assessment tool is accompanied by other tools including a geospatial decision support tool that identifies habitats for mitigation fulfilment and forecasts mitigation costs of proposed developments.
- *Offset siting support tool for the DRECP* (Kreitler et al., 2015[125]): this tool was developed specifically for the area under the Desert Renewable Energy Conservation Plan (DRECP) of California but could be adapted for other offset analyses. Based on the hypothetical impacts from 15 331 ha of solar development in the Western Mojave Desert, the authors applied the tool to compare two offset scenarios. The first scenario prioritised offsets according to impacted features, while the second scenario prioritised offsets to maximize regional biodiversity conservation gains. The two methods only agree on 28% of their prioritized sites and differ in meeting species-specific offset goals. Differences between the two scenarios highlight the importance of clearly specifying choices and priorities for offset siting and mitigation in general.

5.2.2. Government support (e.g., grants and subsidies)

Governments can support biodiversity-aligned deployment of renewable power through grants and subsidies. These could be administered nationally, sub-nationally or internationally (e.g., through official development assistance – see Chapter 4). Government support can be used to 1) enhance knowledge and evidence of renewable energy impacts; 2) promote research, development and demonstration of technologies, decision-support tools and practices for addressing biodiversity impacts of renewable energy developments; and 3) incentivise renewable power companies and utilities to seek positive outcomes for biodiversity above and beyond regulatory requirements. Each of these areas is discussed below and supported by examples.

Enhancing knowledge and evidence of renewable energy impacts

Gaps in data and knowledge of renewable energy impacts on biodiversity remain (see Chapter 3). The evidence differs across technologies (i.e., solar, offshore wind, onshore wind), taxa (e.g., avian impacts are relatively well-studied), ecosystems (e.g., relatively few studies on impacts in marine and forest ecosystems), and geographies (e.g., the evidence base is relatively strong for offshore wind in North Sea compared to Mediterranean). In addition to promoting systematic collection of data from pre-construction biodiversity surveys and post-construction monitoring, governments have a role in promoting targeted research to address knowledge and evidence gaps through grants and by co-ordinating research initiatives. Examples include:

- *Deploying Solar with Wildlife and Ecosystem Services Benefits (SolWEB)*: Through the SolWEB programme, the U.S. Department of Energy is providing USD 14 million to fund research on how solar energy interacts with wildlife and ecosystems. Approximately USD 8 million has been granted to projects focussing on solar-wildlife interaction with projects ranging from quantification of insect biodiversity and pollinator communities at solar facilities through to evaluation of the response of pronghorn and other mammals to utility-scale solar facilities. Approximately USD 5 million has been granted to projects addressing ecosystem services, including the development of a tool to assess ecosystem service benefits provided by utility-scale solar facilities and the development of a national soil data collection system at solar facilities to enable soil health and ecosystem services assessments. The SolWEB projects are part of the D.O.E.'s approximately USD 100 million research portfolio on renewable energy and biodiversity interactions (US DOE, n.d.[126]).
- *Enabling Coexistence Options for Wind Energy and Wildlife (ECO Wind)*: The US National Renewable Energy Laboratory (NREL) has awarded USD 1.1 million to three industry teams to support research on bats and wind energy projects. Wind turbines are a leading cause of bat mortality, but relatively little is understood about how bats interact with wind turbines. The awardees include Bowman and Wildlife Imaging Systems, who will investigate how bats behave around wind turbines in different environments to understand whether they behave differently in different geographic regions; Electric Power Research institute who will investigate whether bats prefer calmer air directly behind wind turbines or turbulent air surrounding them to determine whether better understanding airflow around turbines can help deter bats from approaching them; Stantec Consulting Services will assess where and how bats use airspace near the rotor-swept area to assess differences between technologies, specifically acoustic detectors and cameras, to monitor bats and determine the factors that influence the different species' behaviours.

Promoting RDD for biodiversity-aligned renewable energy development

Through grants, governments can promote research, development and demonstration of technologies, decision-support tools and good practices for reducing harmful biodiversity impacts or promoting positive impacts from renewable power infrastructure. Technologies could include low-risk sources of renewable power or mitigation technologies for existing renewable power sources such as AI bird-identification technology to support shutdown-on-demand of wind turbines. While the focus here is on grants, to optimise innovation governments will need to ensure the broader enabling environment facilitates and promotes biodiversity-aligned innovations. This entails, for example, well-aligned competition, trade and investment policies coupled with strong environmental policy that encourage developers to find low-cost solutions to address biodiversity impacts from renewable energy. Examples of grant or subsidy schemes for promoting RDD for biodiversity-aligned renewable energy include:

- *Offshore Renewable Impacts on Ecosystem Services (ORIES):* UK Research and Innovation (UKRI), a non-departmental public body sponsored by the Department for Business, Energy and Industrial Strategy (BEIS), has awarded Plymouth Marine Laboratory with an approximately GBP 300 000 (~ USD 363 000) grant to fund the development of a decision-support tool for marine wind energy. The tool, ORIES, aims to help stakeholders to understand the impacts of planned offshore wind installations on marine biodiversity and ecosystem services. ORIES is intended to be operational and publicly available in 2023. UKRI's grant has been complemented by funding from Plymouth Marine Laboratory (~GBP 75 000) and a donation from the Garfield Weston Foundation (GBP 40 000) (PLM, 2022[127]).

- *Examples to Accommodate Biodiversity in Nordic Offshore Wind Projects:* Nordic Energy Research, acting on behalf of the Nordic Committee of Senior Officials for Energy Policies, invited all interested parties to submit an offer for a tender for "Examples to Accommodate Biodiversity in Nordic Offshore Wind Projects". The aim of the tender is two-fold: 1) identify good examples of coexistence between offshore wind projects and biodiversity in the Nordic region and neighbouring countries; and 2) assess the feasibility of mitigation measures in a Nordic context, from practical, technical, economic and ecological perspective (Nordic Energy Research, 2021[128]).

Incentivising adoption of biodiversity-friendly practices in construction and operation of renewable energy facilities

While strong regulatory instruments are essential for safeguarding biodiversity in renewable power developments, governments can encourage developers to seek positive outcomes for biodiversity through subsidies. Subsides are still used in many countries to support the deployment of renewable energy (although there is discussion of whether these should be phased out now that renewables are increasingly cost-competitive) (Held et al., 2019[129]; Melliger and Chappin, 2022[130]). These subsidies can be tailored to reflect biodiversity considerations or be complemented by subsidies that incentivise measures to enhance biodiversity outcomes. If well-designed and targeted, subsidies could help deliver on biodiversity and climate objectives (OECD, 2021[131]), and facilitate innovation (Criscuolo et al., 2022[132]). To ensure efficient use of resources, it will be important to evaluate the effectiveness of subsidies over time to ensure they do not become counter-productive, redundant or market-distortive. Ecosystem valuation could help inform the appropriate volume of the incentive to efficiently internalise the ecosystem service benefits (Siegner et al., 2019[133]). Examples of biodiversity-relevant subsidies for renewable power include:

- *Solar Massachusetts Renewable Target (SMART) programme:* In Massachusetts, US, the Solar Massachusetts Renewable Target (SMART) programme promotes solar development by providing tariff-based incentives to operators of eligible solar arrays. The volume of support provided by the government depends on the category of land proposed for projects and other factors. Solar development in areas designated as brownfields, eligible landfills, and "previously developed areas"[7] receive higher payment rates than those in greenfields. Developments in lands designated as "Priority Habitat," "Core Habitat" or "Critical Natural Landscape," as defined in the statute are ineligible. Additionally, the Massachusetts Department of Energy Resources recently proposed a USD 0.0025/kWh rate adder for solar developments that meet the pollinator-friendly standard established by the University of Massachusetts (Commonwealth of Massachusetts, 2020[134]).
- *Roof-mounted solar subsidies, India:* While not explicitly targeting biodiversity, subsidies for solar panels on rooftops could benefit biodiversity as solar panels installed on existing infrastructure reduce the need for ground-mounted power facilities which are more land-use intensive and have higher risks to biodiversity (Kim et al., 2021[135]). India is one example of a country with subsidies for roof-mounted solar (India, 2023[136]). Depending on the state, subsidies amount to 30% or 70% of installation costs.

5.3. Information instruments and voluntary approaches

Information instruments and voluntary approaches form an important complement to regulatory and economic measures. These include biodiversity-explicit power procurement, industry guidelines, and environmental labelling or ecolabelling, each of which is discussed below. Other examples include voluntary corporate commitments (see examples in 5.1.1 and 5.2.1) and investor performance standards (see Box 5.9). Some instruments (e.g., guidelines) serve to facilitate the implementation of regulatory and economic instruments, while others help to fill a gap in regulation or to drive and reward environmental performance that goes above and beyond regulation (e.g., ecolabels). Several biodiversity-relevant information instruments and voluntary approaches are emerging as the renewable energy sector develops. Opportunities exist for scaling these approaches.

5.3.1. Biodiversity-explicit procurement policies, tender processes and power purchase agreements

Through electricity procurement decisions, governments and other electricity off-takers can promote and support the development of renewable power projects that deliver positive outcomes for biodiversity. This can be achieved by incorporating biodiversity considerations in tendering processes and power purchase agreements (PPAs). PPAs are the contracts governing the sale and purchase of electricity. Through PPAs, electricity sellers (e.g., privately-owned power producers) and buyers (e.g., a state-owned electricity utility or private electricity provider) agree on several criteria such as the amount, timing and cost of energy supply, payment terms, penalties and how much electricity is to be sourced from renewables. Buyers could take a more holistic view to "green" energy, requiring electricity producers to demonstrate their commitment not only to renewables, but also their commitment to biodiversity protection.

Biodiversity can be integrated into tendering and power purchase agreements in various ways. For example, biodiversity benefits can be a consideration or explicit criterium for evaluating proposals from competing renewable power producers or operators. Alternatively, attainment of a biodiversity standard or application of a specific practice can be a prerequisite for a proposal to be considered. These approaches are demonstrated in the examples below. The first five examples cover public or private electricity suppliers serving end-users (Netherlands, France, Clean Power Alliance, MCE and Excel Energy), while the sixth example is of a corporate end-user (Salesforce).

- Hollandse Kust West Wind Farm Zone (HKWWFZ) is located approximately 28.6 nautical miles (53 kilometres) off the west coast of the Netherlands. Two wind farm sites are designated within the HKWWFZ: HKW Wind Farm Site VI and VII. The Dutch Government issued tenders for the permits to develop the sites. Applications were assessed on non-price criteria in addition to price criteria. For Site VI, applications were assessed on four criteria: 1) amount of financial offer; 2) certainty of wind farm being completed; 3) contribution to energy supply; and 4) contribution to the ecology of the North Sea. Notably, the biodiversity-specific fourth criteria accounted for 50% of the total points available and was therefore decisive. The criterion was split into two: 1) Stimulation of investments to benefit naturally occurring biodiversity (species, populations, and habitats) in the Dutch North Sea and 2) Stimulation of innovation and development of solutions to benefit naturally occurring biodiversity in the Dutch North Sea from the wind farm at Site VI and future Dutch offshore wind farms (RVO, 2019[137]).
- In France, developers of land-based solar energy projects compete to supply electricity to the national grid. Biodiversity-relevant factors are considered in the tendering process both as a prerequisite for a proposal to be considered and as criteria against which a proposal is scored. To compete, projects must first meet local criteria. For example, facilities not located within an urban plan must have the favourable opinion from the departmental commission regarding the preservation of natural, agricultural and forest areas. In natural areas, sites cannot be in wetlands. Once this prerequisite is met, projects are evaluated against a set of criteria with a total of 100 points available. Nine points focus on the environment. If the area of installation is degraded (e.g., a former industrial site, polluted site or waste area), the project receives the full nine points; otherwise, the project receives zero of these points (France, 2021[138]). For offshore wind

developments, tenders are scored based on the budget they allocate to environmental measurement and a biodiversity fund. Tenders for a 1 GWh project receive a maximum score if they allocate at least EUR 75 million.

- Clean Power Alliance (CPA) is an electricity provider serving South California. It is California's largest local community choice aggregator[8] (Clean Power Alliance, 2022[139]). In 2018, CPA adopted an Environmental Stewardship Principle, committing to provide customers with energy that delivers multiple benefits for nature, air and water. To fulfil this commitment, environmental stewardship is one of six key evaluation criteria used by CPA to assess proposals for renewable energy and storage power purchase agreements. Biodiversity data and information provided by environmental NGOs were used to develop qualitative assessment questions and a qualitative evaluation framework for ranking projects according to environmental stewardship (CPA, 2022[140]; Hughes, 2020[141]).
- MCE is a community choice aggregator operating in 37 communities across four counties in California, US. Since 2020, the electricity provider requires all new ground-mounted solar project partners to plant pollinator-friendly ground cover throughout the facility. In addition, developers must submit a pollinator-friendly solar scorecard within 30 days of the Commercial Operation Date, within two years of Commercial Operation when installation of pollinator habitat must be completed, and then after five, ten and fifteen years of Commercial Operation. The requirement applies to both their Feed-in Tariff programme and power purchase agreements. Additionally, MCE encourages specific solar array design elements to be considered to support pollinator-friendly habitats and reduce maintenance costs. The pollinator-friendly solar scorecard allows points to be awarded for positive planned actions (e.g., planned % of site dominated by native plant species cover) and subtracted for harmful planned actions (e.g., pesticide application).
- Xcel Energy, a utility serving eight states in the United States, plans to add at least 3 000 MW of solar generation by 2030 (Xcel Energy, 2019[142]). In 2018, they announced plans to require all future solar project proposals to disclose information on what type of vegetation will be planted. To facilitate disclosure of this information, developers are required to complete a pollinator habitat scorecard, which was developed by the State of Minnesota. While achievement of the voluntary "pollinator friendly" standard is not a prerequisite, the habitat scorecard enables the criteria to be factored into Excel Energy's decision and signals to developers the importance of ensuring solar facilities are nature-friendly (Act4Nature, 2021[143]; Morehouse, 2018[144]).
- In 2013, Salesforce made a public commitment to reach 100% renewable energy. This involves purchasing renewable energy and certificates equivalent to the amount of power used in their global operations. Salesforce procures renewable energy by evaluating each project against a set of environmental, social and economic attributes collated in a renewable energy project matrix (Lorenzen and Scher, 2018[145]). The attributes cover, among other things, "land use and habitat", and "wildlife". Information gathered during the Request for Proposal process is used to evaluate projects and guide the selection process. For the land use and habitat attribute, points differ depending on the habitat within which the project is sited, ranging from one point for projects sited in natural habitat or prime farmland to five points for projects in built environment (parking lots, brownfields, and rooftops). Owing to the high biodiversity values of critical habitat, Salesforce does not accept projects located in these areas. For the wildlife attribute, points range from one to five, with full points given to projects that go above and beyond standard industry best management practices or voluntarily offset impacts offsite. Projects that fail to assess impacts or respect wildlife regulations are excluded. The project evaluation matrix is accompanied by guidance on how to deliver on the attributes.

As demonstrated by these examples, integrating biodiversity into PPA and tendering processes could be used to demand increasingly higher standards for biodiversity in renewable power projects, requiring companies to go beyond regulation to be competitive. It could also help drive innovation. As with other instruments, ongoing monitoring will be necessary to ensure effective implementation of the criteria as well as penalties to dissuade non-compliance.

Beyond the biodiversity-benefits, integrating biodiversity into tendering processes and PPA could have the added benefit of electricity producers considering the costs of biodiversity management in their tariffs. Financing for a renewable power projects is often secured after a PPA has been signed, which means the environmental policies – and the additional costs of implementing them – are not considered in the PPA and therefore must be subsequently addressed as an add-on cost instead (Hulka and Conzo, 2021[146]).

5.3.2. Industry guidelines on biodiversity and renewable power infrastructure

Voluntary industry guidelines can facilitate mainstreaming of biodiversity in renewable power development. They can achieve this by clearly communicating relevant regulations, providing a framework for evaluating and addressing biodiversity impacts, and presenting good practices and tools for mitigating negative impacts and promoting positive impacts. In addition to improving biodiversity outcomes, adherence to guidelines could help developers reduce the risks of project delays, biodiversity-related liability and penalties.

Industry guidelines on renewable power and biodiversity have been developed by supranational, national and subnational government bodies, non-governmental organisations and advisory firms (Table 5.4), often with wide stakeholder engagement. Examples of industry guidelines produced or endorsed by governments are presented below.

- The European Commission's Guidance Document on Wind Energy Developments and EU Nature Legislation (European Commission, 2020[19]) aims to guide primarily developers, consultants and competent authorities on how best to ensure that wind energy developments are compatible with the EU Birds and Habitats Directives. The guidance covers the pre-construction, construction, operation and decommissioning or repowering phases. It outlines the legislative framework for renewable energy and nature in Europe, provides guidance on screening, assessment and strategic planning, discusses potential effects of wind energy on nature and outlines key considerations for monitoring and adaptive management. Similar guidance on energy transmission infrastructure and nature was published in 2018, targeting project developers, transmission system operators (TSOs) and competent authorities (European Commission, 2018[62]).
- The US Fish and Wildlife Service Land-Based Wind Energy Guidelines (USWFS, 2012[147]) aim to: 1) Promote compliance with relevant wildlife laws and regulations; 2) Encourage scientifically rigorous survey, monitoring, assessment, and research designs proportionate to the risk to species of concern; 3) Produce potentially comparable data across the Nation; 4) Mitigate, including avoid, minimize, and compensate for potential adverse effects on species of concern and their habitats; and 5) Improve the ability to predict and resolve effects locally, regionally and nationally. The guidelines were developed in collaboration with a Wind Turbine Guidelines Advisory Committee, which included representatives from federal energy and wildlife agencies, state energy commissions and wildlife agencies, tribes, renewable power companies, conservation organisations, and academia. Adherence to the Guidelines is voluntary and does not absolve developers of their liability under the Migratory Bird Treaty Act, the Bald and Golden Eagle Protection Act and the Endangered Species Act. However, if developers are found to be in violation of these Acts, the US FSW and Office of Law Enforcement will consider a developer's documented efforts to communicate with the US FWS and adhere to the Guidelines when considering prosecution. Those adhering to the Guidelines are considered to have taken reasonable and effective measures to avoid the "take" of protected species. Supplementary guidance – *Eagle Conservation Plan Guidance* – focussing on bald and golden eagle protection in wind energy development has also been produced by the USFWS (U.S. FWS, 2013[148]). Similar guidelines on

solar power siting do not exist, however, some developers have applied the wind power guidelines to solar projects (TNC, 2022[149]).

The following factors could help ensure the wide uptake and effectiveness of voluntary guidelines. First, engaging a range of stakeholders, including biodiversity experts, industry associations, regulators and civil society groups in the design of the guidelines could help ensure their rigour and relevance. From an ecological perspective, it is critical that the guidelines are evidence-based and updated as evidence evolves. Second, engaging industry associations to disseminate and promote the guidelines throughout the sector and provide training on their application where needed (McKenney, 2020[150]). Third, encouraging public and private companies to report on the extent to which the guidelines have been applied, as part of voluntary or mandatory reporting processes. Fourth, making financial institutions aware of the guidelines and require developers to apply them to receive financing. In the US, for example, some financial institutions are aware that renewable energy developers may incur legal and financial risks related to violating existing laws protecting animals from injury or death. These financial institutions may look more favourably on financing proposals that incorporate USFWS guidelines in their development and operations plans (U.S., 2022[151]).[9] Fifth, national level guidance could be translated at subnational level, to adapt to locally specific regulatory and environmental circumstances. For example, a few US states (e.g., Arizona, Nebraska, Wyoming), have provided state-level wind guidance based on the US Fish and Wildlife Service Land-Based Wind Energy Guidelines.

Table 5.4. Examples of guidance for mainstreaming biodiversity into renewable power infrastructure

Name of guidance	Geographic scope	Energy infrastructure covered	Biodiversity component covered	Target audience	Lead proponent
International guidance					
Renewable energy technologies and migratory species: Guidelines for sustainable deployment	Global	All renewable energy infrastructure	Migratory species	Government and project developers	UNEP CMS
Mitigating biodiversity impacts associated with solar and wind energy	Global	Solar CSP, PV; onshore and offshore wind; and electricity grid network	All components	Industry (e.g., wind/solar energy developers and investors)	TBC and IUCN
Guidelines on How to Avoid or Mitigate Impact of Electricity Power Grids on Migratory Birds in the African-Eurasian Region	Regional - African-Eurasian Region	Electricity grid network	Migratory birds	Governments (Parties to AEWA)	UNEP CMS, AEWA, Raptors MoU
Guidelines for consideration of bats in wind farm projects Revision 2014	Regional – Europe	Onshore and offshore wind energy	Bats	Governments (Parties to EUROBAT)	UNEP EUROBAT (Agreement on the Conservation of Populations of European Bats)
Guidance on Energy Transmission Infrastructure and EU Nature Legislation	Regional – Europe	Electricity grid network	Wildlife and habitats (under EU Birds and Habitats Directives)	Developers, transmission system operators (TSOs) and competent authorities	European Commission
Guidance Document on Wind Energy Developments and EU Nature Legislation	RegionalL – Europe	Onshore and offshore wind energy	Wildlife and habitats (under EU Birds and Habitats Directives)	Developers, consultants and competent authorities	European Commission
National and subnational guidance					

Name of guidance	Geographic scope	Energy infrastructure covered	Biodiversity component covered	Target audience	Lead proponent
The US Fish and Wildlife Service Land-Based Wind Energy Guidelines	United States	Land-based wind energy siting, construction and operation	Wildlife and habitats	Wind energy developers	US FWS
Eagle Conservation Plan Guidance	United States	Wind energy siting, construction and operation	Birds – specifically bald and golden eagles	Wind energy developers	US FWS
Handbook for preventing bird electrocution	Norway	Electricity grid network	Birds	Electricity grid planners and developers	Norwegian Water Resources and Energy Directorate
Guidance on landscaping when building hydropower and energy production facilities	Norway	Wind, power lines and hydropower	Habitats and landscapes	Developer/operators	Norwegian Water Resources and Energy Directorate
Knowledge base on the effects of wind power on land	Norway	Onshore wind	Habitats, species and landscapes. Other social and environmental considerations.	Developers/operators and planners	Norwegian Water Resources and Energy Directorate
South African Good Practice Guidelines for Operational Monitoring for Bats at Wind Energy Facilities	South Africa	Wind	Bats	Wind energy developers and environmental consultants	South African Bat Assessment Association
South African Bat Fatality Threshold Guidelines	South Africa	Wind	Bats	Wind energy developers and environmental consultants	South African Bat Assessment Association
Pollinator Smart: Comprehensive Manual	Virginia, United States	Solar	Species and habitats	Solar energy developers	Virginia Department of Conservation and Recreation; Virginia Department of Environmental Quality
Reducing Impacts to Birds, Bats from Wind Energy	California, United States	Wind	Species	Local permitting agencies	California Energy Commission; California Department of Fish and Wildlife

Source: (Prinsen et al., 2012[59]), Guidelines on How to Avoid or Mitigate Impact of Electricity Power Grids on Migratory Birds in the African-Eurasian Region; (Rodrigues et al., 2014[152]), Guidelines for consideration of bats in wind farm projects Revision 2014; (European Commission, 2018[62]), Guidance on Energy Transmission Infrastructure and EU nature legislation Environment; (European Commission, 2020[19]), Commission notice Guidance document on wind energy developments and EU nature legislation Commission notice Guidance document on wind energy developments and EU nature legislation Guidance document on wind energy developments and EU Nature Legislation; (U.S. FWS, 2013[148]) Eagle Conservation Plan Guidance Module 1-Land-based Wind Energy; (Bennun et al., 2021[80]); Mitigating biodiversity impacts associated with solar and wind energy development: guidelines for project developers; (Aronson et al., 2020[72]), South African Good Practice Guidelines for Operational Monitoring for Bats at Wind Energy Facilities 2nd edition for Operational Monitoring for Bats at Wind Energy Facilities-ed 2; (Macewan et al., 2020[153]); South African Bat Fatality Threshold Guidelines Edition 3 April 2020 (Virginia DCR and Virginia DEQ, 2019[154]), Pollinator-Smart: Comprehensive Manual.

5.3.3. Environmental labelling and information schemes

Environmental labelling and information schemes (ELIS) refer to a broad set of policies and initiatives that provide information to external users about one or more aspects of the environmental performance of a product or service (Prag, Lyon and Russillo, 2016[155]). They may involve either business-to-business communication or business-to-consumer communication. ELIS can be mandatory (e.g., energy efficiency labels in Europe) or voluntary (Prag, Lyon and Russillo, 2016[155]). They can operate at an international, national or subnational level.

Until recently, ELIS in the power sector have focussed on communicating energy efficiency or renewable sourcing of energy. While biodiversity-explicit ELIS have existed for several decades [e.g., Forest Stewardship Council (est. 1992) and the Programme for the Endorsement of Forest Certification (est. 1999)], they are relatively new to the power sector. Examples of operational ELIS in the power sector that include biodiversity criteria are outlined below:

- *EKOenergy Ecolabel:* The EKOenergy ecolabel is an internationally recognised non-for-profit ecolabel for renewable electricity that was established by a network of environmental NGOs. The purpose of the ecolabel is to help energy suppliers sell a recognisable and widely accepted product, increase the positive impact of renewable energy consumption and make it easier for consumers to navigate the energy market and communicate about their purchase. To attain the EKOenergy label, energy must be 100% renewable while also meeting additional environmental sustainability criteria established based on consultation with environmental NGOs, energy companies, consumers, consumer organisations and public authorities.

  To protect biodiversity, wind and solar installations located in a) nature reserves designated by the authorities; b) Natura 2000 areas; c) Important Bird and Biodiversity Areas; d) UNESCO World Heritage Sites are only accepted if the EKOenergy Board approves them, after consultation with relevant stakeholders. For solar energy, this approval can be made if a management plan is implemented that covers elements such as a) fencing (avoiding habitat fragmentation and maximising access for animals); b) pesticide free management; c) measures to avoid land sealing; d) habitat management on the area between the panels and on the unbuilt parts of the sites; e) water management. EKOenergy's Secretariat organises an annual audit to verify that sold/used EKOenergy-labelled volumes fulfil all the requirements listed in EKOenergy's criteria. The audit is based on facts and figures that are certified or confirmed by public authorities and/or reliable third-party certification organisations (EKOenergy, 2021[156]).

- *The Blue Dot Network quality infrastructure certification:* The Blue Dot Network is a certification for infrastructure projects that is being developed by Australia, Japan and the United States. It is largely derived from the G20 Principles for Quality Infrastructure Investment and other international standards such as the International Finance Corporation's Performance Standards (IFC PS), the OECD Guidelines for Multinational Enterprises, and the OECD Recommendation on the Governance of Infrastructure, among others. It applies to infrastructure projects across all major sectors, including renewable energy, in both developed and developing economies. Blue Dot Network element 8 (there are 10 elements in total) *Uphold international best practices for environmental and social safeguards, including respect for labour and human rights*, incorporates criteria relating to biodiversity that reflect IFC PS 6. The certification is being piloted on projects around the world, including onshore and offshore wind farms.

- *Seal of Excellence in Sustainability (Sello de Excelencia Sostenibilidad)*: The Spanish Photovoltaic Union[10] (UNEF) launched a solar PV sustainability certificate in 2021 and awarded its first certificate in December 2021 to Iberdrola's *Renovables' Andévalo* facility. Certification requires fulfilment of criteria in four areas: socio-economic impact, governance, environmental integration and biodiversity protection, and circular economy. The biodiversity criteria include siting the facility outside Natura 2000 Networks; conducting a cumulative impact assessment; using permeable fencing; putting in measures to promote biodiversity such as nesting sites, ponds and insect hotels; prevention of soil degradation and plantation of trees or planting of new trees where transplantation is not possible.

  The certification process is divided into two phases. In the development phase, a preliminary certificate is issued based on the analysis of the project documentation provided by the client. In the second phase, after construction, the final certificate is issued once the on-site evaluation of the plant has been carried out and it is verified that the project has been developed based on the documentation previously provided. If the plant is already built, the documentary analysis and the on-site assessment is carried out at the same time. The certification process is audited by a third-party (UNEF, 2021[157]).

- *Minnesota Habitat Friendly Solar Programme:* In Minnesota, the Board of Soil and Water Resources (BSWR) has launched a Habitat Friendly Solar Programme which provides technical guidance and assessment criteria (BSWR, 2020[158]). Although a voluntary scheme, Minnesota legislative requirements state that "an owner of a solar site implementing solar site management practices may claim that the site provides benefits to gamebirds, songbirds and pollinators only if the site adheres to guidance set forth by the pollinator plan provided by the Board of Water and Soil Resources" (Minnesota, 2021[159]). The legislation intends to prevent "greenwashing" or false environmental claims.

  To meet the standard, projects must complete a Solar Site Pollinator Habitat Assessment scorecard and score 70 points or more. A gold standard is attained with >85 points. Projects must be inspected yearly to identify any management needs and a monitoring form must be completed. At the end of the third year of vegetation establishment for the project, and every three years afterwards, a qualified natural resource staff with plant ID knowledge must fill out an Established Project Assessment Form. Criteria include, for example, the percent of native plant cover and the diversity of plant cover (# of plant species with >1% cover). Points are subtracted for activities such as insecticide use (BSWR, 2020[158]). Preliminary evidence indicates the scheme may benefit flowering plants and insects (Lukens, 2021[160]). At least eight more states in the US (Illinois, Maryland, Massachusetts, Michigan, New York, North Carolina, South Carolina and Vermont) have since established voluntary pollinator-friendly certification programmes (Terry et al., 2020[161]; Dowling, 2020[162]).

As illustrated by these examples, ELIS can be developed and managed by public agencies, private companies and NGOs, either individually or in partnership. Governments can be involved in ELIS in various ways, including by (co-)developing or (co-)managing schemes, and endorsing or incentivising them (e.g., through subsidies, including beneficial tax status – see economic instruments).

The effectiveness of environmental labelling and information schemes may be influenced by various factors including the scientific-robustness and stringency of the criteria, the process for verifying and auditing attainment of criteria, and the enforcement measures in place to address non-compliance and dissuade false environmental claims (Klintman, 2016[163]; OECD, 2013[164]). As ELIS for biodiversity and renewable power develop, it would be beneficial to monitor and evaluate their effectiveness. Building adaptability into ELIS is good practice, so that standards and procedures can be revised and strengthened based on lessons learned and advances in scientific knowledge (OECD, 2013[164]).

References

Act4Nature (2021), *Upscaling Corporate Solutions for Biodiversity*, http://www.act4nature.com/wp-content/uploads/2021/03/upscaling-corporate-solutions-for-biodiversity.pdf (accessed on 4 May 2022). [143]

Amerson, A. et al. (2022), "A Summary of Environmental Monitoring Recommendations for Marine Energy Development That Considers Life Cycle Sustainability", *Journal of Marine Science and Engineering*, Vol. 10/5, p. 586, https://doi.org/10.3390/jmse10050586. [82]

Aronson, J. et al. (2020), *South African Good Practice Guidelines for Operational Monitoring for Bats at Wind Energy Facilities 2nd edition for Operational Monitoring for Bats at Wind Energy Facilities-ed 2*, http://www.sabaa.org.za/documents/SABAA_Operational_Bat_Monitoring_Guidelines_2ndEd_June2020.pdf (accessed on 26 October 2022). [72]

Barr, Z. et al. (2021), "The Permitting, Licensing and Environmental Compliance Process: Lessons and Experiences within U.S. Marine Renewable Energy", *Energies*, Vol. 14/16, p. 5048, https://doi.org/10.3390/en14165048. [32]

Bas, A. and L. Dieckhoff (2021), *La compensation ecologique pour les projet eoliens terrestres et photovoltaiques au sol en France et en Allemange*, Office franco-allemand pour la transition energetique (OFATE), https://energie-fr-de.eu/files/ofaenr/04-notes-de-synthese/02-acces-libre/06-societe-environnement-economie/2021/OFATE_EIFER_Synthese_Compensation_Ecologique_2104.pdf (accessed on 5 May 2022). [111]

BBOP (2009), *Business and Biodiversity Offsets Programme: Compensatory Conservation Case Studies*, http://www.forest-trends.org/biodiversityoffsetprogram/guidelines/non-bbop-case-studies.pdf. (accessed on 28 April 2022). [108]

Bennett, G., M. Gallant and K. Ten Kate (2017), *State of Biodiversity Mitigation 2017: Markets and Compensation for Global Infrastructure Development*, http://www.forest-trends.org/wp-content/uploads/2018/01/doc_5707.pdf. [104]

Bennun, L. et al. (2021), *Mitigating biodiversity impacts associated with solar and wind energy development: guidelines for project developers*, IUCN, International Union for Conservation of Nature, https://doi.org/10.2305/iucn.ch.2021.04.en. [80]

Bern Convention (2004), *Recommendation No. 110 (2004) on minimising adverse effects of above-ground electricity transmission facilities (power lines) on birds*. [58]

Brazil (2022), *Response to OECD Questionnaire on Biodiversity and Renewable Energy*. [88]

BSWR (2020), "Habitat Friendly Solar Site Assessment Form for Project Planning", https://bwsr.state.mn.us/sites/default/files/2020-05/Habitat%20Friendly%20Solar%20Site%20Assessment%20Form%20for%20Project%20Planning%205-26-2020.pdf (accessed on 14 October 2022). [158]

Bull, J. and N. Strange (2018), "The global extent of biodiversity offset implementation under no net loss policies", *Nature Sustainability*, Vol. 1/12, pp. 790-798, https://doi.org/10.1038/s41893-018-0176-z. [116]

Butterworth, T. et al. (2019), "Biodiversity net gain. Good practice principles for development Case studies", *CIRIA C776b*. [107]

California Energy Commission (2022), *Building Energy Efficiency Standards*, https://www.energy.ca.gov/programs-and-topics/programs/building-energy-efficiency-standards/2022-building-energy-efficiency (accessed on 24 February 2023). [71]

Caro-Gonzalez, A., J. Toro and M. Zamorano (2021), "Effectiveness of environmental impact statement methods: A Colombian case study", *Journal of Environmental Management*, Vol. 300, p. 113659, https://doi.org/10.1016/j.jenvman.2021.113659. [48]

CBD (2012), *DECISION ADOPTED BY THE CONFERENCE OF THE PARTIES TO THE CONVENTION ON BIOLOGICAL DIVERSITY AT ITS ELEVENTH MEETING XI/18. Marine and coastal biodiversity: sustainable fisheries and addressing adverse impacts of human activities, voluntary guidelines for environmental assessment, and marine spatial planning A. Addressing biodiversity considerations in fisheries management, and addressing adverse impacts of human activities on marine and coastal biodiversity*, https://www.cbd.int/doc/decisions/cop-11/cop-11-dec-18-en.pdf (accessed on 11 July 2022). [42]

CBD (2006), *Impact assessment: Voluntary guidelines on biodiversity-inclusive impact assessment. COP 8 Decision viii/28*, https://www.cbd.int/decision/cop/?id=11042. [41]

Chambert, T. and A. Besnard (2021), "Assessing the demographic impact of bird collisions with wind turbines. State of the art and methodological recommendations", https://mape.cnrs.fr/wp-content/uploads/2022/04/Briefing-note_WP2_Assessing-the-demographic-impact-of-bird-collisions-with-wind-turbines.pdf (accessed on 3 November 2022). [13]

Clean Power Alliance (2022), *About Us - Clean Power Alliance*, https://cleanpoweralliance.org/about-us/ (accessed on 4 March 2022). [139]

CMS (2020), *Reconciling Energy Development with the Conservation of Migratory Species: an Analysis of National Reports to CMS COP13*, https://www.cms.int/en/document/reconciling-energy-development-conservation-migratory-species-analysis-national-reports-c-0. [40]

Commonwealth of Massachusetts (2020), *Commonwealth of Massachusetts Executive Office of Energy and Environmental Affairs: Solar Massachusetts Renewable Target Programme: Guideline*, https://www.mass.gov/doc/land-use-and-siting-guideline-october-2020/download (accessed on 17 October 2022). [134]

COWI (2009), *Study concerning the report on the application and effectiveness of the EIA Directive Final report*, https://ec.europa.eu/environment/archives/eia/pdf/eia_study_june_09.pdf (accessed on 12 July 2022). [89]

CPA (2022), *Request for Offer (RFO) - Clean Power Alliance*, https://cleanpoweralliance.org/request-for-offer-rfo/ (accessed on 5 May 2022). [140]

Criscuolo, C. et al. (2022), "An industrial policy framework for OECD countries: Old debates, new perspectives", *OECD Science, Technology and Industry Policy Papers*, No. 127, OECD Publishing, Paris, https://doi.org/10.1787/0002217c-en. [132]

Croll, D. et al. (2022), "Framework for assessing and mitigating the impacts of offshore wind energy development on marine birds", *Biological Conservation*, Vol. 276, p. 109795, https://doi.org/10.1016/j.biocon.2022.109795. [119]

CSBI (2015), *cross-sector guide for implementing the mitigation hierarchy.*, repared by the Biodiversity Consultancy on behalf of IPIECA, ICMM and the Equator Principles Association: Cambridge UK., http://www.csbi.org.uk/our-work/mitigation-hierarchy-guide/ (accessed on 12 July 2022). [123]

Deparment of Environmental Affairs (n.d.), *EIA Guideline for Renewable Energy Projects. Department of Environmental Affairs*, https://www.dffe.gov.za/sites/default/files/legislations/EIA_guidelineforrenewableenergyprojects_0.pdf. [50]

Dowling, Z. (2020), "Developing a Pollinator-friendly Certification Program for Sola Photovoltaic Arrays in Massachusetts", https://ag.umass.edu/sites/ag.umass.edu/files/news/pdf/dev_pollinator-friendly_pv_certif_prog.pdf (accessed on 14 October 2022). [162]

Durning, B. and M. Broderick (2018), "Development of cumulative impact assessment guidelines for offshore wind farms and evaluation of use in project making", *Impact Assessment and Project Appraisal*, Vol. 37/2, pp. 124-138, https://doi.org/10.1080/14615517.2018.1498186. [52]

Dutta, N. et al. (2021), "From Project Impacts to Strategic Decisions: Recurring Issues and Concerns in Wind Energy Environmental Assessments", *Environmental Management*, Vol. 68/4, pp. 591-603, https://doi.org/10.1007/s00267-021-01518-2. [46]

EC (2022), *Accelerating permitting for renewable energy*, https://reform-support.ec.europa.eu/accelerating-permitting-renewable-energy_en. [23]

EC (2022), *Guidance to Member States on good practices to speed up permit-granting procedures. SWD(2022) 149 final*, https://eur-lex.europa.eu/legal-content/EN/TXT/PDF/?uri=CELEX:52022SC0149&from=EN. [31]

EEB (2022), *Top 10 Problems for Renewable Energy in Europe are not linked with Nature Protection*, https://eeb.org/wp-content/uploads/2022/05/Top-10-problems-for-renewable-energy-in-Europe.pdf. [27]

EKOenergy (2021), *EKOenergy-Governance structure and criteria for electricity*, http://www.isealalliance.org. (accessed on 13 October 2022). [156]

Energy Transitions Commission (2023), *Solution Toolkit: Actions for national/regional governments and policymakers*, https://www.energy-transitions.org/wp-content/uploads/2023/01/Barriers_PP_GovernmentST_vFinal.pdf. [30]

Energy Transitions Commission (2023), *Streamlining planning and permitting to accelerate wind and solar deployment*, https://www.energy-transitions.org/wp-content/uploads/2023/01/Barriers_PlanningAndPermitting_vFinal.pdf. [6]

Equator Principles Limited (2022), *The Equator Principles*, https://equator-principles.com/about-the-equator-principles/ (accessed on 25 October 2022). [100]

EU (2022), *COMMISSION DELEGATED REGULATION (EU) 2022/1288*, https://eur-lex.europa.eu/legal-content/EN/TXT/PDF/?uri=CELEX:32022R1288&from=EN. [87]

European Commission (2022), *REPowerEU Plan*, https://eur-lex.europa.eu/legal-content/EN/TXT/?uri=COM%3A2022%3A230%3AFIN&qid=1653033742483. [24]

European Commission (2020), *Commission notice Guidance document on wind energy developments and EU nature legislation Commission notice Guidance document on wind energy developments and EU nature legislation Guidance document on wind energy developments and EU Nature Legislation*, European Commission (C2020) 7730 final, Brussels, https://ec.europa.eu/environment/nature/natura2000/management/docs/wind_farms_en.pdf (accessed on 3 May 2022). [19]

European Commission (2018), *Guidance on Energy Transmission Infrastructure and EU nature legislation Environment*, https://doi.org/10.2779/827210. [62]

France (2021), *Cahier des charges de l'appel d'offres portant sur la réalisation et l'exploitation d'Installations de production d'électricité à partir de l'énergie solaire « Centrales au sol »*, https://www.cre.fr/media/Fichiers/publications/appelsoffres/2021-pv-sol-telecharger-le-cahier-des-charges-en-vigueur. [138]

France Treasury (2021), *Decrees, orders, circulars: Decree no. 2021-663 of 27 May 2021 implementing Article L.533-22-1 of the Monetary and Financial Code*, https://www.tresor.economie.gouv.fr/Articles/80af1116-2fcd-47d0-ad1d-ea24352e6295/files/273f9026-bbc4-4fc2-ba60-f86f6fe16c1f (accessed on 6 February 2023). [86]

FWS, U. (n.d.), *Eagle Management*, https://www.fws.gov/program/eagle-management/eagle-permits. [11]

Gill, E. and C. Hein (2022), *IEA Wind White Paper Cumulative Effects Analysis for Wind Energy Development: Current Practices, Challenges, and Opportunities*, IEA Wind TCP Task 34. Prepared by National Renewable Energy Laboratory (US) for the International Energy Agency (IEA), http://www.nrel.gov/publications. (accessed on 11 July 2022). [51]

Government of France (2011), *Installation of ground-mounted photovoltaics: Guide for impact assessments*, https://www.ecologie.gouv.fr/sites/default/files/Guide_EI_Installations-photovolt-au-sol_DEF_19-04-11.pdf (accessed on 17 June 2022). [49]

Government of Ireland (2022), *Planning permission exemptions for rooftop solar panels on homes and other buildings proposed*, https://www.gov.ie/en/press-release/3b9bf-planning-permission-exemptions-for-rooftop-solar-panels-on-homes-and-other-buildings-proposed/#. [9]

Gullen Range Wind Farm (2016), *Project Approvals and Documents |*, https://gullenrangewindfarm.com/the-project/project-approvals-and-documents/ (accessed on 3 February 2023). [85]

Hayes, M. et al. (2019), "A smart curtailment approach for reducing bat fatalities and curtailment time at wind energy facilities", *Ecological Applications*, Vol. 29/4, https://doi.org/10.1002/eap.1881. [67]

Held, A. et al. (2019), "Do Almost Mature Renewable Energy Technologies Still Need Dedicated Support Towards 2030?", *Economics of Energy & Environmental Policy*, Vol. 8/2, https://doi.org/10.5547/2160-5890.8.2.ahel. [129]

Hooper, T., M. Austen and A. Lannin (2021), "Developing policy and practice for marine net gain", *Journal of Environmental Management*, Vol. 277, p. 111387, https://doi.org/10.1016/j.jenvman.2020.111387. [118]

Howard (2018), *Industry evidence programme: offshore windfarms - pilot industry evidence base. Report to The Crown Estate and Royal HaskoningDHV.* [90]

Hughes, P. (2020), *Integrating Environmental Stewardship Principles into Power Procurement*, Clean Power Exchange, https://cleanpowerexchange.org/integrating-environmental-stewardship-principles-into-power-procurement/ (accessed on 4 March 2022). [141]

Hulka, S. and L. Conzo (2021), *Recommendations for integrating biodiversity into the wind energy sector in emerging market countries*, https://blog.private-sector-and-development.com/2021/06/15/recommendations-for-integrating-biodiversity-into-the-wind-energy-sector-in-emerging-market-countries/. [146]

Iberdrola (2023), *Our goal: to have a net positive impact on biodiversity by 2030*, https://www.iberdrola.com/sustainability/environment/iberdrola-biodiversity#:~:text=Iberdrola%20will%20have%20a%20net,quantify%2C%20and%20monitor%20compliance%20therewith. [37]

IEA (2022), *Renewable Energy Market Update*, https://www.iea.org/reports/renewable-energy-market-update-may-2022. [21]

IEA (2022), *Renewables 2022*, https://www.iea.org/reports/renewables-2022. [22]

IEA (2018), *20 Renewable Energy Policy Recommendations*, https://www.iea.org/reports/20-renewable-energy-policy-recommendations. [33]

IFC (2017), *Tafila Region Wind Power Projects Cumulative Effects Assessment*, International Finance Corporation, World Bank Group, http://www.ifc.org (accessed on 1 July 2022). [57]

IFC (2012), "International Finance Corporation's Guidance Note 6: Biodiversity Conservation and Sustainable Management of Living Natural Resources Disclaimer Guidance Note 6 Biodiversity Conservation and Sustainable Management of Living Natural Resources", http://www.millenniumassessment.org. (accessed on 5 October 2021). [105]

IFC (2012), *Performance Standards on Environmental and Social Sustainability.* [99]

India, G. (2023), *National Portal for Rooftop Solar - Ministry of New and Renewable Energy*, https://solarrooftop.gov.in/ (accessed on 7 February 2023). [136]

Indonesia, R. (2012), *Ministry of Environment Regulation 5 2012 on Types of Business/Activities that Require EIA.* [20]

Ireland EPA (2022), *Strategic Environmental Assessment | Environmental Protection Agency*, SEA Spatial Information Sources Inventory, https://www.epa.ie/publications/monitoring--assessment/assessment/strategic-environmental-assessment/sea-spatial-information-sources-inventory-.php (accessed on 15 July 2022). [91]

IRENA and IEA-PVPS (2016), *End-Of-Life Management: Solar Photovoltaic Panels*, https://www.irena.org/publications/2016/Jun/End-of-life-management-Solar-Photovoltaic-Panels. [1]

IUCN, TBC and DICE (2021), *Global Inventory of Biodiversity Offset Policies (GIBOP). World View - A snapshot of national biodiversity offset policies*, https://portals.iucn.org/offsetpolicy/. [106]

Jack, T. (2022), "Offshore Wind Energy Permitting Processes in the European Union: An examination of Danish, German, Scottish and Swedish offshore permitting processes and case study of acoustic impact on marine mammals". [26]

Jacob, C. et al. (2020), "Marine biodiversity offsets: Pragmatic approaches toward better conservation outcomes", *Conservation Letters*, Vol. 13/3, https://doi.org/10.1111/conl.12711. [117]

Jenkins, A. et al. (2015), *Birds and Wind-Energy Best-Practice Guidelines for assessing and monitoring the impact of wind-energy facilities on birds in southern Africa*. [73]

Josimović, B., A. Cvjetić and D. Furundžić (2021), "Strategic Environmental Assessment and the precautionary principle in the spatial planning of wind farms – European experience in Serbia", *Renewable and Sustainable Energy Reviews*, Vol. 136, p. 110459, https://doi.org/10.1016/j.rser.2020.110459. [44]

Kettel, E. et al. (2022), "Better utilisation and transparency of bird data collected by powerline companies", *Journal of Environmental Management*, Vol. 302, p. 114063, https://doi.org/10.1016/j.jenvman.2021.114063. [92]

Kim, J. et al. (2021), "Current site planning of medium to large solar power systems accelerates the loss of the remaining semi-natural and agricultural habitats", *Science of The Total Environment*, Vol. 779, p. 146475, https://doi.org/10.1016/J.SCITOTENV.2021.146475. [55]

Kim, J. et al. (2021), "Current site planning of medium to large solar power systems accelerates the loss of the remaining semi-natural and agricultural habitats", *Science of The Total Environment*, Vol. 779, p. 146475, https://doi.org/10.1016/j.scitotenv.2021.146475. [135]

Klintman, M. (2016), "A Review of Public Policies relating to the Use of Environmental Labelling and Information Schemes (ELIS)", *OECD Environment Working Papers*, No. 105, OECD Publishing, Paris, https://doi.org/10.1787/5jm0p34bk7hb-en. [163]

Le Maître, J, K. et al. (2017), *Bat Mortality Caused by Wind Turbines: Review of impacts and mitigation measures*. [65]

Longcore, T. (ed.) (2015), "Conservation Planning for Offsetting the Impacts of Development: A Case Study of Biodiversity and Renewable Energy in the Mojave Desert", *PLOS ONE*, Vol. 10/11, p. e0140226, https://doi.org/10.1371/journal.pone.0140226. [125]

Lorenzen, M. and M. Scher (2018), *More than a Megawatt: Embedding Social and Environmental Impact in the Renewable Energy Procurement Process*, Salesforce, https://c1.sfdcstatic.com/content/dam/web/en_us/www/assets/pdf/sustainability/sustainability-more-than-megawatt.pdf (accessed on 23 June 2022). [145]

Lukens, L. (2021), "Monitoring Pollinators on Minnesota Solar Installations", https://monarchjointventure.org/images/uploads/documents/Monitoring_Pollinators_on_Minnesota_Solar_Installations_Final_Report_MJV.pdf (accessed on 14 October 2022). [160]

Macewan, K. et al. (2020), "South African Bat Fatality Threshold Guidelines Edition 3 April 2020", http://pza.sanbi.org/vegetation (accessed on 26 October 2022). [153]

Maron, M. et al. (2015), "Conservation: Stop misuse of biodiversity offsets", *Nature*, Vol. 523/7561, pp. 401-403, https://doi.org/10.1038/523401a. [121]

May, J., R. Hobbs and L. Valentine (2017), "Are offsets effective? An evaluation of recent environmental offsets in Western Australia", *Biological Conservation*, Vol. 206, pp. 249-257, https://doi.org/10.1016/j.biocon.2016.11.038. [120]

McClure, C. et al. (2022), "Confirmation that eagle fatalities can be reduced by automated curtailment of wind turbines", *Ecological Solutions and Evidence*, Vol. 3/3, https://doi.org/10.1002/2688-8319.12173. [68]

McKenney, B. (2020), *Clean and Green Pathways for the Global Renewable Energy Buildout*. [150]

Melliger, M. and E. Chappin (2022), "Phasing out support schemes for renewables in neighbouring countries: An agent-based model with investment preferences", *Applied Energy*, Vol. 305, p. 117959, https://doi.org/10.1016/j.apenergy.2021.117959. [130]

Minnesota (2021), *2021 Minnesota Statutes: 216B.1642 Solar Site Management*, https://www.revisor.mn.gov/statutes/cite/216B.1642 (accessed on 14 October 2022). [159]

MoE Saskatchewan (2018), *Adaptive Management Guidelines for Saskatchewan Wind Energy Projects*, https://publications.saskatchewan.ca/#/products/90235. [75]

Morehouse, C. (2018), *In bid to help bees, Xcel to require vegetation disclosure in solar RFPs | Utility Dive*, INDUSTRY DIVE, https://www.utilitydive.com/news/in-bid-to-help-bees-xcel-to-require-vegetation-disclosure-in-solar-rfps/539521/ (accessed on 5 May 2022). [144]

MTES (2018), *Protocole de suivi environnemental des parcs éoliens terrestres*. [78]

MTES (n.d.), *The Multiannual Energy Plan*, Ministere de la Transition Ecologique et Solidaire, https://www.ecologie.gouv.fr/sites/default/files/4pages_PPE_GB_DEF_Web.pdf (accessed on 23 February 2023). [43]

NatureScot (2021), *Guidance - Assessing the cumulative landscape and visual impact of onshore wind energy developments | NatureScot*, https://www.nature.scot/doc/guidance-assessing-cumulative-landscape-and-visual-impact-onshore-wind-energy-developments (accessed on 12 July 2022). [53]

NatureScot (2018), *Guidance - Assessing the cumulative impacts of onshore wind farms on birds | NatureScot*, https://www.nature.scot/doc/guidance-assessing-cumulative-impacts-onshore-wind-farms-birds (accessed on 12 July 2022). [54]

New York State (2016), *Guidelines for Conducting Bird and Bat Studies at Commercial Wind Energy Projects*, https://tethys.pnnl.gov/sites/default/files/publications/NYSDEC-2016.pdf. [74]

ngh environmental (2013), *Biodiversity Offset Management Plan: Broken Hill Solar Plant*, https://www.agl.com.au/-/media/aglmedia/documents/about-agl/how-we-source-energy/broken-hill/20130814-broken-hill-solar-biodiversity-offset-management-plan.pdf?la=en&hash=1646AB4E3CBF5DF70AA717BAE30D1C25 (accessed on 29 April 2022). [109]

Niner, H. et al. (2021), "Exploring the practical implementation of marine biodiversity offsetting in Australia", *Journal of Environmental Management*, Vol. 295, p. 113062, https://doi.org/10.1016/j.jenvman.2021.113062. [113]

Niner, H. et al. (2017), "A global snapshot of marine biodiversity offsetting policy", *Marine Policy*, Vol. 81, pp. 368-374, https://doi.org/10.1016/j.marpol.2017.04.005. [114]

Nordic Energy Research (2021), *Examples to Accommodate Biodiversity in Nordic Offshore Wind Projects-Invitation to tender*, https://www.nordicenergy.org/wordpress/wp-content/uploads/2021/03/Invitation-to-tender_Examples-to-Accommodate-Biodiversity-in-Nordic-Offshore-Wind-Projects_2021.pdf (accessed on 7 February 2023). [128]

NVE (2023), *Licensing of solar power plants*, https://www.nve.no/konsesjon/konsesjonsbehandling-av-solkraftverk/. [8]

NVE (2022), *Concession*, https://www.nve.no/konsesjon/. [15]

Nwanekezie, K., B. Noble and G. Poelzer (2022), "Strategic assessment for energy transitions: A case study of renewable energy development in Saskatchewan, Canada", *Environmental Impact Assessment Review*, Vol. 92, p. 106688, https://doi.org/10.1016/j.eiar.2021.106688. [45]

NYSERD (2021), *Wildlife Data Standardization and Sharing: Environmental Data Transparency for New York State Offshore Wind Energy*, Report prepared by Jenkins, E. and Williams, K. Biodiversity Research Institute, Portland, ME. [95]

OECD (2022), *OECD Environmental Performance Reviews: Norway 2022*, OECD Environmental Performance Reviews, OECD Publishing, Paris, https://doi.org/10.1787/59e71c13-en. [16]

OECD (2022), *Responsible Business Conduct Due Diligence for Project and Asset Finance*, https://www.oecd-ilibrary.org/finance-and-investment/responsible-business-conduct-due-diligence-for-project-and-asset-finance-transactions_952805e9-en. [97]

OECD (2022), *Towards a More Resource-Efficient and Circular Economy: The Role of G20*, http://OECD-G20-Towards-a-more-Resource-Efficient-and-Circular-Economy.pdf. [2]

OECD (2021), "Biodiversity, natural capital and the economy: A policy guide for finance, economic and environment ministers", *OECD Environment Policy Papers*, No. 26, OECD Publishing, Paris, https://doi.org/10.1787/1a1ae114-en. [131]

OECD (2021), *OECD Environmental Performance Reviews: Finland 2021*, OECD Environmental Performance Reviews, OECD Publishing, Paris, https://doi.org/10.1787/d73547b7-en. [17]

OECD (2021), *OECD Environmental Performance Reviews: Ireland 2021*, OECD Environmental Performance Reviews, OECD Publishing, Paris, https://doi.org/10.1787/9ef10b4f-en. [18]

OECD (2020), "Improving resource efficiency and the circularity of economies for a greener world", *OECD Environment Policy Papers*, No. 20, OECD Publishing, Paris, https://doi.org/10.1787/1b38a38f-en. [3]

OECD (2018), *OECD Due Diligence Guidance for Responsible Business Conduct*, https://mneguidelines.oecd.org/OECD-Due-Diligence-Guidance-for-Responsible-Business-Conduct.pdf. [96]

OECD (2017), *MNE Guidelines*, http://mneguidelines.oecd.org/database/instances/nl0032.htm. [102]

OECD (2016), *Biodiversity Offsets: Effective Design and Implementation*, OECD Publishing, Paris, https://doi.org/10.1787/9789264222519-en. [103]

OECD (2016), *Extended Producer Responsibility: Updated Guidance for Efficient Waste Management*, OECD Publishing, Paris, https://doi.org/10.1787/9789264256385-en. [4]

OECD (2013), *Scaling-up Finance Mechanisms for Biodiversity*, OECD Publishing, Paris, https://doi.org/10.1787/9789264193833-en. [164]

OECD (2012), *MNE Guidelines*. [101]

OECD (2012), *OECD Environmental Outlook to 2050: The Consequences of Inaction*, OECD Publishing, Paris, https://doi.org/10.1787/9789264122246-en. [5]

OECD (2011), *OECD Guidelines for Multinational Enterprises, 2011 Edition*, OECD Publishing, Paris, https://doi.org/10.1787/9789264115415-en. [98]

OECD (1993), *OECD Core Set of Indicators for Environmental Performance Reviews*. [81]

Orsted (2022), *Restoring biodiversity around the Humber Estuary*, https://orsted.com/en/sustainability/our-priorities/nature/north-east-lincolnshire-biodiversity-restoration. [38]

Peplinski, W. et al. (2021), "Marine Energy Environmental Permitting and Compliance Costs", *Energies*, Vol. 14/16, p. 4719, https://doi.org/10.3390/en14164719. [36]

PLM (2022), *PML awarded grant to facilitate expansion of environmentally-sustainable marine UK wind energy | Plymouth Marine Laboratory*, https://www.pml.ac.uk/News/PML-awarded-grant-to-facilitate-expansion-of-envir (accessed on 24 October 2022). [127]

Pollard, E. and L. Bennun (2016), "Who are Biodiversity and Ecosystem Services Stakeholders?", *Day 3 Wed, April 13, 2016*, https://doi.org/10.2118/179458-ms. [35]

Prag, A., T. Lyon and A. Russillo (2016), "Multiplication of Environmental Labelling and Information Schemes (ELIS): Implications for Environment and Trade", *OECD Environment Working Papers*, No. 106, OECD Publishing, Paris, https://doi.org/10.1787/5jm0p33z27wf-en. [155]

Primmer, E. et al. (2019), "Institutions for governing biodiversity offsetting: An analysis of rights and responsibilities", *Land Use Policy*, Vol. 81, pp. 776-784, https://doi.org/10.1016/j.landusepol.2018.11.040. [122]

Prinsen, H. et al. (2012), *Guidelines on How to Avoid or Mitigate Impact of Electricity Power Grids on Migratory Birds in the African-Eurasian Region*, AEWA Conservation Guidelines No. 14, CMS Technical Series No. 29, AEWA Technical Series No. 50, CMS aptors MOU Technical Series No. 3, https://www.unep-aewa.org/sites/default/files/publication/ts50_electr_guidelines_03122014.pdf (accessed on 29 July 2022). [59]

Raptor Protection of Slovakia (2019), *Protecting birds from power lines focusing on countries of Danube/Carpathian region*, https://energy.danube-region.eu/wp-content/uploads/sites/6/sites/6/2019/09/Study_Birds_FIN.pdf (accessed on 7 February 2023). [60]

RBINS (2023), *Management of the marine environment by the MUMM Scientific Surface: Wind Farms in the North Sea*, https://odnature.naturalsciences.be/mumm/en/windfarms/. [83]

Rodrigues, L. et al. (2014), *Guidelines for consideration of bats in wind farm projects Revision 2014*, EUROBATS, https://www.eurobats.org/sites/default/files/documents/publications/publication_series/pubseries_no6_english.pdf (accessed on 29 July 2022). [152]

ROSA (2021), *Offshore Wind Energy Project Monitoring Framework and Guidelines*, https://tethys.pnnl.gov/sites/default/files/publications/ROSA_offshore_wind_project_monitoring_framework_guidelines.pdf. [76]

RVO (2019), *Windenergiegebied Hollandse Kust (west)*, https://www.rvo.nl/onderwerpen/windenergie-op-zee/hollandse-kust-west. [137]

SCBD (2004), *Akwé: Kon Guidelines*, https://www.cbd.int/doc/publications/akwe-brochure-en.pdf. [56]

Schippers, P. et al. (2020), “Mortality limits used in wind energy impact assessment underestimate impacts of wind farms on bird populations”, *Ecology and Evolution*, Vol. 10/13, pp. 6274-6287, https://doi.org/10.1002/ece3.6360. [14]

Shaffer, J., C. Loesch and D. Buhl (2019), “Estimating offsets for avian displacement effects of anthropogenic impacts”, *Ecological Applications*, Vol. 29/8, https://doi.org/10.1002/eap.1983. [124]

Sheppard, J. et al. (2015), “An autonomous GPS geofence alert system to curtail avian fatalities at wind farms”, *Animal Biotelemetry*, Vol. 3/1, https://doi.org/10.1186/s40317-015-0087-y. [69]

Shumway, N. et al. (2018), “The Risks and Opportunities of Translating Terrestrial Biodiversity Offsets to the Marine Realm”, *BioScience*, Vol. 68/2, pp. 125-133, https://doi.org/10.1093/biosci/bix150. [115]

Siegner, K. et al. (2019), *Maximizing land-use benefits from utility scale solar: A cost-benefit analysis of pollinator-friendly solar in Minnesota*, Yale Center for Business and the Environment, https://cbey.yale.edu/sites/default/files/2019-12/MaximizingLandUseBenefitsFromUtility-ScaleSolar_0.pdf (accessed on 6 October 2021). [133]

Sielicki, J. (2020), *Quick Guidance for Preventing Electrocution Impacts on Birds. Reference note. International Association for Falconry and Conservation of Birds of Prey.*, https://birdelectrocution.org/quick-guidance-preventing-electrocution-on-birds/. [63]

Spain (2008), *Royal Decree 1432/2008*, https://www.miteco.gob.es/es/biodiversidad/temas/conservacion-de-especies/especies-silvestres/tendidos/ce-silvestres-tendidos-RD.aspx. [61]

Squires, K. et al. (2021), “Timing and Weather Offer Alternative Mitigation Strategies for Lowering Bat Mortality at Wind Energy Facilities in Ontario”, *Animals*, Vol. 11/12, p. 3503, https://doi.org/10.3390/ani11123503. [66]

Stephenson, P. (2021), *A Review of Biodiversity Data Needs and Monitoring Protocols for the Off shore Wind Energy Sector*, https://renewables-grid.eu/fileadmin/user_upload/_RGI_Report_PJ-Stephenson_October.pdf (accessed on 24 February 2023). [77]

Sud, R. and S. Patnaik (2022), *How does permitting for clean energy infrastructure work?*, https://www.brookings.edu/research/how-does-permitting-for-clean-energy-infrastructure-work/. [7]

Susskind, L. et al. (2022), "Sources of opposition to renewable energy projects in the United States", *Energy Policy*, Vol. 165, p. 112922, https://doi.org/10.1016/j.enpol.2022.112922. [34]

Tebb, J. (2018), *AGL Broken Hill Solar Plant Biodiversity Offset Site Ecological Monitoring Report-Broken Hill Offset Site IA170600*, Jacobs Australia Pty Limited, http://www.jacobs.com (accessed on 29 April 2022). [110]

Terry, G. et al. (2020), "State Pollinator-Friendly Solar Initiatives", https://www.cesa.org/wp-content/uploads/State-Pollinator-Friendly-Solar-Initiatives.pdf. [161]

The Crown Estate (2023), *Marine Data Exchange*, https://www.marinedataexchange.co.uk/. [93]

TNC (2022), *Site Renewables Right: Accelerating a Clean and Green Renewable Energy Buildout in the Central United States*, The Nature Conservancy, https://www.nature.org/content/dam/tnc/nature/en/documents/SRR_Methods_20220202_LR.pdf (accessed on 10 August 2022). [149]

Tougaard, J., K. Beedholm and P. Madsen (2022), "Thresholds for noise induced hearing loss in harbor porpoises and phocid seals", *The Journal of the Acoustical Society of America*, Vol. 151/6, pp. 4252-4263, https://doi.org/10.1121/10.0011560. [70]

Transalta (2022), *Wolfe Island - TransAlta*, https://transalta.com/about-us/our-operations/facilities/wolfe-island/ (accessed on 3 February 2023). [84]

U.S. (2022), *United States Response to OECD Working Party on Biodiversity Water and Ecosystems Questionnaire*. [151]

U.S. FWS (2013), *Eagle Conservation Plan Guidance Module 1-Land-based Wind Energy*, U.S. Fish and Wildlife Service. [148]

UK (2021), *Environment Act 2021*, https://www.legislation.gov.uk/ukpga/2021/30/schedule/14/enacted. [12]

Ulibarri, N., B. Cain and N. Ajami (2017), "A Framework for Building Efficient Environmental Permitting Processes", *Sustainability*, Vol. 9/2, p. 180, https://doi.org/10.3390/su9020180. [28]

UN Environment (2018), *Assessing Environmental Impacts - A Global Review of Legislation | Capacity4dev*, UN Environment, https://europa.eu/capacity4dev/unep/documents/assessing-environmental-impacts-global-review-legislation (accessed on 5 April 2022). [39]

Underwood, E., K. Taylor and G. Tucker (2018), "The use of biodiversity data in spatial planning and impact assessment in Europe", *Research Ideas and Outcomes*, Vol. 4, p. e28045, https://doi.org/10.3897/rio.4.e28045. [94]

UNEF (2021), *Sello de Excelencia en Sostenibilidad | UNEF*, https://www.unef.es/es/sello-sostenibilidad (accessed on 14 October 2022). [157]

US DOE (n.d.), *DOE Invests $14 Million to Enhance Environmental and Wildlife Benefits from Solar Energy Infrastructure | Department of Energy*, https://www.energy.gov/articles/doe-invests-14-million-enhance-environmental-and-wildlife-benefits-solar-energy (accessed on 24 October 2022). [126]

US FWS (2021), *Federal Register :: Updated Collision Risk Model Priors for Estimating Eagle Fatalities at Wind Energy Facilities*, https://www.federalregister.gov/documents/2021/05/05/2021-09362/updated-collision-risk-model-priors-for-estimating-eagle-fatalities-at-wind-energy-facilities (accessed on 27 October 2022). [10]

USWFS (2012), *U.S. Fish and Wildlife Service Land-Based Wind Energy Guidelines*, https://www.fws.gov/midwest/es/wind/guidance.html. [147]

Vaissière, A. et al. (2014), "Biodiversity offsets for offshore wind farm projects: The current situation in Europe", *Marine Policy*, Vol. 48, pp. 172-183, https://doi.org/10.1016/j.marpol.2014.03.023. [112]

Vasconcelos, R. et al. (2022), "Environmental licensing for offshore wind farms: Guidelines and policy implications for new markets", *Energy Policy*, Vol. 171, p. 113248, https://doi.org/10.1016/j.enpol.2022.113248. [29]

Virginia DCR and Virginia DEQ (2019), *Pollinator-Smart: Comprehensive Manual*, https://www.dcr.virginia.gov/natural-heritage/document/solar-site-comprehensive-manual.pdf. [154]

Whitby, M., M. Schirmacher and W. Frick (2021), *The State of the Science on Operational Minimization to Reduce Bat Fatality at Wind Energy Facilities*. [64]

White House (2022), *The Biden-Harris Permitting Action Plan to Rebuild America's Infrastructure, Accelerate the Clean Energy Transition, Revitalize Communities, And Create Jobs*, https://www.whitehouse.gov/wp-content/uploads/2022/05/Biden-Harris-Permitting-Action-Plan.pdf (accessed on 9 February 2023). [25]

Wilding, T. et al. (2017), "Turning off the DRIP ('Data-rich, information-poor') – rationalising monitoring with a focus on marine renewable energy developments and the benthos", *Renewable and Sustainable Energy Reviews*, Vol. 74, pp. 848-859, https://doi.org/10.1016/j.rser.2017.03.013. [79]

Xcel Energy (2019), *2020-2034 Upper Midwest Integrated Resource Plan*. [142]

Zhang, J., P. Christensen and L. Kørnøv (2013), "Review of critical factors for SEA implementation", *Environmental Impact Assessment Review*, Vol. 38, pp. 88-98, https://doi.org/10.1016/j.eiar.2012.06.004. [47]

Notes

[1] Examples can be found in the Netherlands and Denmark, among others. See Netherlands Enterprise Agency (2022), Dutch Offshore Wind Guide and Eclareon (2021), Technical support for RES policy development and implementation – Denmark.

[2] Ninety countries submitted national reports.

[3] It is unknown to what extent renewable energy is covered in these countries' legislation.

[4] An updated version of the principles is available here: https://www.environment.nsw.gov.au/topics/animals-and-plants/conservation-programs/nsw-biodiversity-offsets-policy-for-major-projects/principles-for-use-of-biodiversity-offsets-in-nsw

5 Not specific to renewable energy developments.

6 For a discussion of key design features for effective biodiversity offsets see (OECD, 2016[103]).

[7]Defined as areas "with pre-existing paving, construction, or altered landscapes and does not include altered landscapes resulting from current agricultural use, forestry, or use as preserved natural area."

[8] Community choice aggregation policies enable local entities to aggregate electricity contracts within a specific jurisdiction to procure electricity as a group, rather than individuals.

[9] The extent to which financial institutions systematically evaluate application of best practices is unknown.

[10] The Spanish Photovoltaic Union (UNEF) was founded on May 16, 2012 after the merger of three national photovoltaic associations: the Photovoltaic Business Association (AEF), the Photovoltaic Section of the Association of Renewable Energy Producers (APPA Fotovoltaica) and the Photovoltaic Industry Association (ASIF).

Annex A. International agreements relevant to mainstreaming biodiversity into renewable energy infrastructure

2030 Agenda for Sustainable Development and its Sustainable Development Goals

Achieving sustainable development requires governments to address multiple and interlinked policy challenges, including improving energy access and security, mitigating and adapting to climate change, and reversing biodiversity loss. These policy challenges are embodied in the United Nations (UN) 2030 Agenda for Sustainable Development adopted by UN members in 2015, which lays out 17 global sustainable development goals (SDGs), and a plan to achieve them. The SDGs are "inter-related" and "indivisible"; achieving sustainable development requires governments to deliver across all 17 goals. As potential synergies and trade-offs exist across the SDGs, effective implementation of the 2030 agenda demands policy coordination and coherence. Five SDGs are particularly relevant to mainstreaming biodiversity into the energy sector:

- SDG 7. Ensure access to affordable, reliable, sustainable and modern energy for all, which includes a target (7.2) to increase substantially the share of renewable energy in the final energy consumption,
- SDG 9. Build resilient infrastructure, promote inclusive and sustainable industrialization and foster innovation,
- SDG 13. Take urgent action to combat climate change and its impacts,
- SDG 14. Conserve and sustainably use the oceans, seas and marine resources for sustainable development,
- SDG 15. Protect, restore and promote sustainable use of terrestrial ecosystems, sustainably manage forests, combat desertification, and halt and reverse land degradation and halt biodiversity.

United Nations Framework Convention on Climate Change and the Paris Agreement

With the adoption of the Paris Agreement in 2015, governments committed to hold the increase in global average temperature to well below 2°C above pre-industrial levels and pursue efforts to limit the temperature increase to 1.5°C above pre-industrial levels. To achieve this long-term temperature goal, countries aim to reach global peaking of greenhouse gas emissions as soon as possible to achieve a climate neutral world by mid-century. In their Nationally Determined Contributions, countries communicate the actions they will take to reduce their greenhouse gas emissions in line with these goals. Countries are also invited to submit long-term low greenhouse gas emission development strategies, which provide a longer-term horizon to NDCs, in line with achieving climate neutrality by mid-century.

With electricity, heat and transport accounting for just under three quarters of global greenhouse gas emissions (WRI, 2020[1]), transforming the energy sector is pivotal for achieving the Paris Agreement goals. A large increase in renewable energy infrastructure, primarily for electricity generation, is a core part of this transformation. However, at the same time, countries note in the Preamble of the Paris Agreement "the importance of ensuring the integrity of all ecosystems, including oceans, and the protection of biodiversity [...] when taking action to address climate change". Efforts are therefore required to ensure renewable energy expansion is conducted in a way that protects biodiversity.

G20, G7 and the Quality Infrastructure agenda

Increasing attention on the international policy agenda is given to the importance of not just increasing the quantity of infrastructure investment, but also its quality. In 2019, the Osaka Leaders' Declaration of the Group of Twenty (G20) endorsed a set of six voluntary Principles for Quality Infrastructure Investment to provide the G20 with common strategic direction and aspiration (Group of Twenty, 2019[2]). The underlying aim is "to maximize the positive economic, environmental, social, and development impact of infrastructure and create a virtuous circle of economic activities, while ensuring sound public finances." Principle 3 Integrating Environmental Considerations in Infrastructure Investments is of particular relevance to mainstreaming biodiversity into infrastructure. The principle states that:

> *Both positive and negative impacts of infrastructure projects on ecosystems, biodiversity, climate, weather and the use of resources should be internalized by incorporating these environmental considerations over the entire process of infrastructure investment, including by improving disclosure of these environment related information, and thereby enabling the use of green finance instruments.*

The G20 principles followed the adoption of the five Ise-Shima Principles for Promoting Quality Infrastructure Investment by the Group of Seven (G7) in 2016 (G7, 2016[3]). Principle 3 of the Ise-Shima Principles is Addressing social and environmental impacts emphasises that quality infrastructure investment must consider and address social and environmental impacts of infrastructure projects, including through environmental safeguards that are in line with international best practices.

The OECD works with the G7, G20 and OECD members to support the implementation of Quality Infrastructure. For example, the OECD Compendium of Policy Good Practices for Quality Infrastructure Investment (OECD, 2020[4]) compiles and provides a unique set of existing integrated and multidisciplinary international good practices and measures relevant to policy makers and practitioners to pursue quality infrastructure investment. These good practices promote a shared understanding of the elements needed to support quality infrastructure investments in alignment with the G20 Principles for Quality Infrastructure Investment and in accordance with international standards.

The Compendium is complemented by the OECD Implementation Handbook for Quality Infrastructure Investment (OECD, 2021[5]). While the Compendium is a policy guidance tool, the Handbook is an analytical and operational tool, focussing on selected major issues and challenges, including integration of environmental considerations.

OECD Council Recommendation on the Governance of Infrastructure

The OECD Council Recommendation on Governance of Infrastructure (OECD, 2020[6]) was adopted by the OECD Council in 2020. It is a non-binding legal instrument that provides a tool to assist governments to invest in infrastructure projects in a way that is cost effective, affordable and trusted by investors, citizens and other stakeholders. The recommendation highlights the need for a long-term strategic vision for infrastructure that considers international commitments on environmental protection and the need for environmental considerations in project appraisal, environmental impact assessments and stakeholder engagement that ensures debate and oversight on main environmental impacts.

OECD Council Recommendation concerning the Reduction of Environmental Impacts from Energy Production and Use

The OECD Council Recommendation concerning the Reduction of Environmental Impacts from Energy Production and Use (OECD, 1976[7])was adopted in 1976 and remains in force today. While some of the recommendations are specific to fossil fuels, many of them are equally relevant to renewable energy infrastructure. For example, the Council

> *Recommends that Member countries, in the planning and implementation of their energy and environment policies, ensure that: i) Environment policies and energy policies are integrated, both at the formulation stage and the implementation stage; ii) The public is objectively informed and its views are sought; iii) Land use planning is employed, which takes into account environmental protection goals.*

Convention on Biological Diversity and the Post-2020 Global Biodiversity Framework

The UN Convention on Biological Diversity (CBD) (est. 1992) has three objectives: the conservation of biological diversity, the sustainable use of the components of biological diversity and the fair and equitable sharing of the benefits arising out of the utilization of genetic resources. Achieving these objectives requires biodiversity to be mainstreamed across various policy areas, which is reflected in Articles' 6 (b), 10 (a) (c), 14, 11, 7 (c) and 8 (l) of the Convention Text and the 2011-2020 Strategic Plan for Biodiversity and its Aichi Targets under the Convention on Biological Diversity (CBD), particularly targets under Strategic Goal A.

Subsequent decisions on mainstreaming were adopted by the Conference of the Parties to the CBD, notably Decisions XIII/31 and XIV/32. Decision XIV/3, focuses on the energy sector and infrastructure (in addition to mining, manufacturing and processing sectors), inviting governments as well as public and private entities to include approaches to conserve biodiversity in investment decisions, for example, through strategic environmental assessments and integrated spatial planning, and to review and use existing policies and tools to promote biodiversity-related sustainable production and consumption in these sectors. The decision also establishes a long-term strategic approach to mainstreaming (LTAM) and an informal advisory group, of which OECD is a member, to advise on the further development of the LTAM proposal.

Despite the emphasis given to mainstreaming and progress made by some countries, biodiversity mainstreaming remains insufficient to halt and reverse current trends in biodiversity loss (Diaz, S. et al., 2019[8]). None of the twenty Aichi Targets were fully achieved by 2020. While six targets were partially achieved, none of them were the mainstreaming targets (Aichi Targets 1-4) (SCBD, 2020[9]). Increasing efforts to mainstream biodiversity will therefore be vital for delivering on goals and targets of the Kunming-Montreal Global Biodiversity Framework.

Convention on the Conservation of Migratory Species of Wild Animals (CMS)

Signatories to the Convention on the Conservation of Migratory Species of Wild Animals (CMS), also known as the Bern Convention, have adopted a number of Resolutions directly relevant to mainstreaming biodiversity into energy infrastructure, focusing in particular on addressing the potential impacts on migratory species. These include Resolutions 7.4 Electrocution of Migratory Species, 7.5 Wind Turbines and Migratory Species, 10.11 Power Lines and Migratory Species, 11.27 Renewable Energy and Migratory Species 12.21 Climate Change and Migratory Species.

Resolution 11.27 endorsed guidelines on Renewable Energy Technologies and Migratory Species: Guidelines for Sustainable Development, and urges Parties and non-Parties to the CMS to implement them. Additionally, the Resolution instructed the Secretariat to convene a multi-stakeholder Task Force on Reconciling Selected Energy Sector Developments with Migratory Species Conservation (the Energy Task Force) to support and promote the implementation of the guidance and decisions, and complement them as required. Relevant guidelines and resolutions have also been adopted by other CMS Agreements and Memorandums of Understanding, such as the Agreement on the Conservation of Cetaceans of the Black Sea, Mediterranean Sea and Contiguous Atlantic Area (ACCOBAMS), the Agreement on the Conservation of Small Cetaceans of the Baltic, North East Atlantic, Irish and North Seas (ASCOBANS), the Agreement on the Conservation of Populations of European Bats (EUROBATS), and the Agreement on the Conservation of African-Eurasian Migratory Waterbirds (AEWA).

Ramsar Convention on Wetlands

The Ramsar Convention is the intergovernmental treaty that provides the framework for the conservation and wise use of wetlands and their resource. Signatories to the convention adopted Decision XI.10 Wetlands and energy issues, which covers both renewable and non-renewable sources of energy (COP11 Ramsar, 2012[10]). The Decision was adopted in recognition that some energy activities have direct and indirect adverse impacts on the ecological character of wetlands. Among other things, the Decision invites contracting parties to develop ecological impact criteria to inform energy generate site selection relating to wetlands and to apply such criteria, for example as part of Strategic Environmental Assessments, to guide energy development. The Decision also endorses annexed Guidance for addressing the implications for wetlands of policies, plans and activities in the energy sector.

References

COP11 Ramsar (2012), *Resolution XI.10 Wetlands and Energy Issues*, https://www.ramsar.org/sites/default/files/documents/pdf/cop11/res/cop11-res10-e.pdf (accessed on 19 July 2022). [10]

Diaz, S. et al. (2019), *Summary for policymakers of the global assessment report on biodiversity and ecosystem services of the Intergovernmental Science-Policy Platform on Biodiversity and Ecosystem Services*, IPBES, Bonn, https://www.ipbes.net/system/tdf/ipbes_7_10_add-1-_advance_0.pdf?file=1&type=node&id=35245 (accessed on 6 September 2019). [8]

G7 (2016), *G7 Ise-Shima Principles for Promoting Quality Infrastructure Investment*, https://www.mofa.go.jp/files/000160272.pdf. [3]

Group of Twenty (2019), *G20 Osaka Leaders Declaration*, https://www.mofa.go.jp/policy/economy/g20_summit/osaka19/en/documents/final_g20_osaka_leaders_declaration.html#:~:text=We%2C%20the%20Leaders%20of%20the,for%20the%20benefit%20of%20all. [2]

OECD (2021), *OECD Implementation Handbook for Quality Infrastructure Investment*, https://www.oecd.org/finance/oecd-implementation-handbook-for-quality-infrastructure-investment.htm. [5]

OECD (2020), *OECD Compendium of Policy Good Practices for Quality Infrastructure Investment*. [4]

OECD (2020), *Recommendation of the Council on the Governance of Infrastructure OECD/Legal/0460*, https://legalinstruments.oecd.org/en/instruments/OECD-LEGAL-0460 (accessed on 21 September 2021). [6]

OECD (1976), *Recommendation of the Counicl concerning the Reduciton of Environmental Impacts from Energy Production and Use*, https://legalinstruments.oecd.org/en/instruments/OECD-LEGAL-0149. [7]

SCBD (2020), *Global Biodiversity Outlook 5*. [9]

WRI (2020), *Climate Watch, The World Resources Institute*, http://cait.wri.org/. [1]

Notes

[1] Strategic actions to enhance the implementation of the Strategic Plan for Biodiversity 2011-2020 and the achievement of the Aichi Biodiversity Targets, including with respect to mainstreaming and the integration of biodiversity within and across sector.

[2] Mainstreaming of biodiversity in the energy and mining, infrastructure, manufacturing and processing sectors.

www.ingramcontent.com/pod-product-compliance
Lightning Source LLC
Chambersburg PA
CBHW061213240926
51846CB00006B/1211
* 9 7 8 9 2 6 4 9 6 5 7 9 9 *